The User-Computer Interface in Process Control

A Human Factors Engineering Handbook

The User-Computer Interface in Process Control

A Human Factors Engineering Handbook

Walter E. Gilmore
David I. Gertman
Harold S. Blackman

EG&G Idaho
Idaho National Engineering Laboratory
Idaho Falls, Idaho

ACADEMIC PRESS
Harcourt Brace Jovanovich, Publishers
Boston San Diego New York
Berkeley London Sydney
Tokyo Toronto

ACADEMIC PRESS, INC.
1250 Sixth Avenue, San Diego, CA 92101

United Kingdom Edition published by
ACADEMIC PRESS INC. (LONDON) LTD.
24-28 Oval Road, London NW1 7DX

Library of Congress Cataloging-in-Publication Data

Gilmore, Walter E.
The user-computer interface in process control: a human factors engineering handbook / Walter E. Gilmore, David I. Gertman, Harold S. Blackman

p. cm.
Bibliography: p.
Includes index.
ISBN 0-12-283965-X
1. Process control--Data processing. 2. Human-Computer interaction.
I. Gertman, David. II. Blackman, Harold S. III. Title.
TS156.8M32 1989 89-32045
629.8'9--dc20 CIP

Printed in the United States of America

89 90 91 92 9 8 7 5 4 3 2 1

CONTENTS

ACKNOWLEDGMENTS

In July 1985, the Idaho National Engineering Laboratory (INEL) was asked by the U. S. Nuclear Regulatory Commission (USNRC) to prepare a handbook of human engineering guidelines for assessing user-computer interfaces in nuclear power plants. The resulting technical report (NUREG/CR-4227) became a popular reference for systems designers and specialists from a variety of disciplines. The document provided a compendium of information to support informed decisions about the design of the user-computer interface. From this early effort, the authors were encouraged to take the original material a step further to develop a handbook of guidelines that would be useful to anyone designing computer interfaces in process control settings. This is that further development.

The authors must, at this point, acknowledge deep indebtedness to James P. Jenkins, National Aeronautics and Space Administration (formerly with the USNRC), for his technical guidance and sponsorship of the original report. In addition, we are also indebted to William W. Banks of the Lawrence Livermore National Laboratory (LLNL) for providing the opportunity to pursue this work. We owe a special thanks to Donald L. Schurman of the Idaho National Engineering Laboratory (INEL) and Hugh David of Eurocontrol Centre Experimental for encouraging us to publish this handbook. We would also like to thank Donald L. Schurman for his technical review of the numerous draft versions of this handbook and Kelly Cook for her patience and professionalism in assisting us in typing those drafts. A great deal of credit for the art work goes to Gay Gilbert, and her contributions are also appreciated. Finally, a special thanks is due to the many professionals whose contributions to the field of user-computer interface are referenced throughout this handbook. Without them, this book would not have been possible.

PREFACE

The popularity of computers as integrated components of process control systems is worldwide. Existing nuclear power plants and those under construction are implementing computer-based information presentation. There is a growing trend to provide process control-room operators with video display unit (VDU)-based information. Operators often control all plant processes based on this information. In the nuclear industry, for example, most plants use safety parameter display systems (SPDSs). Long racks of hardwired analog meters and controls, which historically constituted the mainstay of operator control, are no longer so common. Richard Dallimonti[a] has noted, "The introduction of digital control systems based on microprocessors and serial data highway communication brought with it the Cathode-Ray-Tube (CRT) console [VDU] as the new operator's 'window' to the process. As a result, hundreds of control centers have since been designed without the traditional panels full of individualized instruments–instead they are replaced by multiple CRT consoles [VDUs]." Some experts predict that 10% of all workplaces will be equipped with VDU systems in the near future.[b] However, this figure may be somewhat conservative.

There are many reasons for the widespread presence of VDUs in all areas of industry. There are assertions that higher efficiency results from replacing conventional means of display and control with VDUs. A significant decrease in computer hardware costs, coupled with increasingly sophisticated display/input devices, provides another reason to use VDUs. The VDU has emerged as a cheap, powerful, and flexible tool for control room applications. Unfortunately, human factors considerations for providing the user

---

[a] R. Dallimonti, "Challenge for the 80s: Making Man-Machine Interfaces[33] More Effective," Control Engineering, 29, 2, January 1982, pp. 26-30.

[b] E. Grandjean and E. Vigliani (eds.), "Ergonomic Aspects of Visual Display Terminals," Proceedings of the International Workshop, Milan, March 1980, London: Taylor & Francis Ltd., 1983.[60]

with the best possible interface have lagged behind the installation of these systems. In the past, human factors engineering has not been a formal requirement in the product life cycle of VDUs for nuclear, fossil fuel, or chemical plant control room design. In the near term, it still may not be. Thus, there is the need for a number of disciplines, such as computer specialists, design engineers, architectural engineers, and operations personnel, to take an active role in filling voids where the application of human factors guidelines has not been an explicit requirement.

For the sake of efficiency, reviews of existing guidelines are needed, and to merge them into a single reference for designers and evaluators is required. Those who review process control systems cannot afford to wait until all human interface issues have been resolved and the ultimate guidelines documents are written. Computer systems are being installed now in all areas of business and industry. One can assume that established human engineering guidelines will not be incorporated into all of these systems in the final design. There is, then, an immediate demand for a handbook to be used as a tool in the assessment of those systems currently in place. We hope that this handbook, in some small part, can assist in meeting this need.

Purpose and Scope

The purpose of this handbook is to integrate and provide, in a single source, essential human factors guidelines for visual display use and design in process control applications. Many of these guidelines are contradictory in their design recommendations. We offer no apology for this; we are simply presenting the state of knowledge. For example, the recommendations for symbol contrast ratios are highly variable and depend on the source which the user chooses for obtaining the information. We determined that this text would not be complete if the guidelines failed to reflect alternate criteria or solutions.

The strength and validity of many of the guidelines presented here are variable. Some of the guidelines are supported by formal experimental research. Other guidelines are the result of long-standing conventions and lack formal test validation. Many of the latter are the result of standing practices based on research from hardwired analog displays and signs. Extrapolation of this early research to CRT monitors may not apply. When compared with other sciences, human factors engineering is a relatively new discipline; and the existing guidelines are far from complete. Therefore, the gaps in our knowledge must occasionally be filled by best guesses or conventions from similar research.

Recommended Use of This Handbook

This book is intended to serve as an aid to persons tasked with designing, reviewing, assessing, or evaluating a user-computer interface that is currently in place or is proposed for installation in a process control system. Potential users of this handbook may represent a number of various disciplines of engineering and technical areas. They may or may not possess specialized skills in human factors or computer systems. Our intent is to provide all of these people with an overview of systems principles, followed by supporting human factors guidelines.

LIST OF ABBREVIATIONS

AFAAMRL	Air Force Armstrong Aero Medical Research Laboratory
ANS	American Nuclear Society
ANSI	American National Standards Institute
ASHRAE	American Society of Heating, Refrigeration, and Air Conditioning
C	Celsius
C^3	Command, Control, and Communication
cd	candela
cm	centimeter
CRT	Cathode Ray Tube
cfm	cubic feet per minute
dB	decibel
dB(A)	decibel, A-weighted
deg	degree
DVA	Dynamic Visual Acuity
EPRI	Electric Power Research Institute
ENS	European Nuclear Society
F	Fahrenheit
fc	foot-candle
ft	feet
ftL	foot-lambert
fpm	feet per minute
HVAC	heating, ventilation, and air conditioning
HFS	Human Factors Society
Hz	Hertz
INEL	Idaho National Engineering Laboratory
in	inch
IEEE	Institute of Electrical and Electronic Engineers
ISA	Instrument Society of America
K	Kelvin
LOFT	Loss-of-Fluid Test
lx	lux
PDP	Programmable Display Pushbutton
UCI	User Computer Interface
LED	light-emitting diode

m	meter
mL	millilambert
mm	millimeter
min	minute
ms	millisecond
nm	nanometer
NISO	National Information Standards Organization
OAET	Operator Action Event Tree
P-T	pressure-temperature
s	second
SME	Subject Matter Expert
SPDS	Safety Parameter Display System
MRS	Multidimensional Rating Scale
Kg	Kilogram
lb	pound
P&ID	Piping and Instrumentation Diagram
RPM	Revolutions Per Minute
VDT	Visual Display Terminal
VDU	Video Display Unit

I

INTRODUCTION

Background

In recent years, process control, utility, and military industries have witnessed a technical revolution in the use of computerized applications for operating complex process control systems. Previously, the primary concern of control room designers was to be sure that the control room had sufficient space within which to pack the numerous controls and displays required for successful operation. Increased safety requirements and plant complexity have markedly increased the number of controls and displays with which an operator is forced to deal. This human factors "nightmare" became dramatically evident to the general public in the incident at Three Mile Island. A series of human errors, brought on by overburdened operators in a poorly designed control room, forced the nuclear industry to seriously rethink how control rooms should be designed. Now over 50% of the public utilities in the U.S. have submitted to formal control room design reviews (Schurman)[127]. A similar concern over risk in the chemical and manufacturing industry has followed. In the 1980s, more attention than ever before has been given to ensuring that a plant operator has the necessary procedures, training, and equipment to successfully handle a plant emergency.

The introduction of cost-effective computer systems has been instrumental in changing the entire philosophy for designing control rooms. Instead of walking down long rows of meters and switches, as shown in Figure 1, the operator can now operate the plant from one or more centrally located CRT consoles (see Figure 2). The integration of multiple plant parameters into a single set of colorgraphics displays eliminates the time-consuming task of getting this information from panels dispersed throughout the control room. Many of the operator's

Figure 1. The "old-style" process control room. The plant instrumentation and control systems are operated from several upright panels.

Figure 2. A modern computerized process control room. The operators' duties are performed from a CRT console.

duties have now been automated. Computerized process control places the operator in a supervisory capacity, where human intervention rarely takes place, other than during an unusual or off-normal event.

The benefits of computer technology, as with any innovation, are not without their drawbacks. All too often the computer-user interface does not adhere to accepted human factors principles and guidelines. Thus, the introduction of this new technology that was intended to improve operator performance may actually degrade it. An interface that is inadequate because of the absence of labeling, conflicts in coding schemes, or poor user dialogue, reduces the operators' acceptance of a new system. If an operator sees a computerized display system as being a hindrance, he or she will either ignore the system or, in some situations, simply turn it off. In many systems, the human factors considerations are all too often not addressed. In part, this is because designers have not had convenient access to guidelines. This text has been written to introduce software engineers, system designers, and project managers to the principles and practice of human factors engineering in the design, development, and acquisition of computer systems for process control.

This text can best be described as a resource document. It is designed to supply guidance for (a) identifying human factors engineering guidelines and criteria and applying them to the system specification; (b) assuring that human factors engineering considerations have been addressed in the design process; and (c) implementing a human factors engineering program in project planning and development.

What Is Human Factors Engineering?

Human factors engineering is the "discipline concerned with designing equipment so that people can use [it] effectively and safely, and creating environments suitable for human living and work [67]." Stated more simply, it is design based on human characteristics.

Human factors engineering applies what is known about human characteristics and behavior to equipment facilities and environments in a systematic manner. The human factors engineer becomes involved at all places in the system where the user (or operator) must come in contact with the hardware and/or software of a system. It is at these critical junctions between human and machine where the user interface is located. This is the place where the human factors engineer is best suited to apply his/her skills and knowledge to the design of the overall system. Without adequate consideration of the user interface, the resulting system specification is incomplete, and the final design will be driven by software or hardware requirements with little consideration for the requirements of the user. The relationship between these various elements is illustrated in Figure 3. Note that Figure 3 represents the ideal that should be adhered to between the three elements of system hardware, software, and user needs. When the elements are successfully integrated into a system, the final product requires less adaption on the part of the user. This simply translates to a more reliable man-machine system.

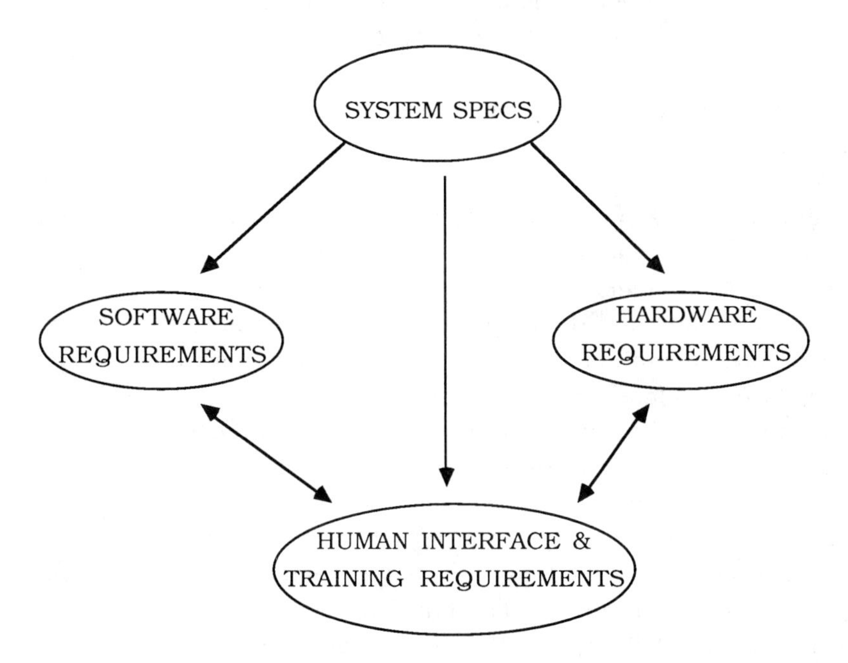

Figure 3. Application of human factors engineering to systems development. Human factors engineering must be applied to properly and completely decompose system specifications. Otherwise, a hardware- or software-driven system is created rather than a user-driven one.

The benefits that can be realized by applying human factors engineering to the design of a computer-assisted process control system include

- improved human performance, shown by increased speed and accuracy and less energy expenditure and fatigue;
- better training and reduced training cost;
- improved use of manpower through minimized need for special skills and aptitudes;
- reduced lost–time and equipment losses related to accidents caused by human error;
- improved comfort, acceptance, and reliability by the user/operator; and
- improved system safety.

The Role of Human Factors Engineering in Computerized Process Control

Wickens [159] has characterized process control as an area in which the issues of decision making, perception, and memory take on importance. He contrasts this with the areas where human performance may be grossly characterized by motor skills. For persons designing equipment or information-processing and presentation systems, this characterization takes on particular significance. Interface design must support the supervisory role and system documentation demands that operators assume in their attempts to regulate complex processes. Furthermore, the relatively high amount of monitoring activity and relatively occasional intervention in plant processes, that constitute the process operator's job is unlike the demands placed upon the ordinary white-collar office worker or upon the journeyman or apprentice welder.

The need to attend to the unique demands of the process control environment and its distributed processing architecture have been noted by Dallimonti [33]. As he stated, "the introduction of digital control systems brought with it the CRT console." Designers shifted from designing traditional panels of instruments, where we have well-defined guidelines, to relatively uncharted waters regarding ways to organize both data and data retrieval via the CRT screen. This evolution in plant technology has given rise to the need for explicit human factors guidelines. To many of us in the human factors engineering profession, it appears that advanced process control technology has been introduced unsystematically into the control room and that science has tried to play catch-up (in terms of conducting experiments and issuing guidelines just prior to disaster). Only within the past decade have guidelines for alternate input devices, such as light pens, trackerballs, mouse, joystick, soft programmable function keys, voice input, and touch screen, emerged. Each type of device, not surprisingly, has certain advantages and disadvantages. To the dismay of some of us, much of the necessary definitive research is yet to be performed for these devices. This is the case despite the fact that the media are flexible enough to allow for experimentation. There are similar needs for research that defines the limits for display formats and media, virtual displays, heads-up displays, and tactile displays for the handicapped.

Wickens [159] notes three distinguishing characteristics of process control:

1. Process control variables which are regulated or controlled are relatively slow in comparison to quickly changing variables such as those found in pilot flight profiles. The time horizon may be one of minutes or hours rather than split seconds.

2. The variables controlled are analog and continuous.

3. The process usually consists of a large number of interrelated variables.

It is fair to say that operators and supervisors of plant processes have an important role to play insofar as prevention, detection, and corrective strategies are concerned. The degree to which their response is successful is, in turn, a function of the man-machine interface, the operator's training, the operator's experience, operator burden, quality of procedures, and the plant design offered the operator.

One of the best collections of basic readings on the operator in process control is to be found in Edwards and Lees [43]. It is interesting to note that out of some twenty-six studies cited, two pertain to nuclear reactors, five to chemical plants, and four to paper mills. In the last ten years, the American Nuclear Society (ANS), the Institute for Electrical and Electronics Engineers (IEEE), and the European Nuclear Society (ENS) have all conducted special topical meetings to explore the impact of human factors and human factors research findings upon efficient plant design and safe power plant operations.

One of the early, yet definitive, works on the nature of the operator in process control was that of Crossman [32], who noted that the emphasis in the process control operator's job was shifting toward the ability to take in information, organize it, and interpret it for action. The goal of the operator, he determined, is fivefold: (1) to regulate or stabilize the process; (2) to adjust the process in order to optimize; (3) to make changes from one product to another; (4) to avoid breakdowns; and (5) after breakdown, to regain normal running of plant as soon as possible. To these five goals, the operator brings the skills of sensing, perceiving, prediction, familiarity with controls, and decision-making to bear. In addition to control skills, the communication skills in information logging, group decision-making, and passing along information to the following shift, play a large part in determining the role of the process operator.

The process operator's goals and skills identified by Crossman have relevance for the modern process control room. The mission of human engineering guidelines for process control should be to (a) support the operator's immediate goals for controlling plant process and (b) support the skills the operator brings to bear in order to run the plant in a safe and efficient manner. The introduction of computerized information presentation systems and computerized operator support systems should meet these criteria.

Problems of User-Interface Design

There are many factors which contribute to deficient user-interface design. Some of the most frequent causes for such failures are described below:

- Predetermined design convention – Often it is the conventions of prior design guidance and operating practices that dictate the requirements for a system. These conventions are often found when a noncomputerized control room is being upgraded with CRT workstations. The resulting displays, generated on the CRTs, are often a software replication of the meters and dials from the original control panels. An example, often seen in "upgraded" control rooms, is the placement of pen and ink strip charts in a softcoded version directly on the CRT screen. The implementation of these concepts, in itself, may not be all bad. The problem resides in the designer's failure to rethink what the best mode could be for presenting information to the operator.

- Isolating the designer from the end user – The end user is seldom brought into the design process. A software engineer, with little or no background in plant operations, is designated to develop the applications for the operator to use. The software engineer is usually given only cursory requirements, with little or no input from the operators. The result consists of an interface that is difficult, if not impossible, to operate.

- Poorly defined functional objectives – The objectives (what is to be accomplished with the introduction of the new system) are often not clearly defined. System objectives are either vague or nonexistent. If specific objectives are not included in the conceptual phase of system development, system designers will rapidly lose sight of what the system is supposed to do.

Most, if not all, of these factors are the result of individual attitude and organization [65]. Designers forget for whom the system is intended. When problems arise, there is a fall-back position: the operator is an adaptable human being and will eventually muddle through.

From the organizational perspective, a nonsystematic approach to the design effort creates an undesirable condition. Without a level of systematic coordination, an atmosphere of limited communication and limited cooperation between organizations is created.

It would be idealistic to think that all the factors that contribute to ineffective design can be completely eradicated. Rather, it should be a concerted effort on the part of the systems engineer and project management to admit, first, that problems in interface design can occur in spite of one's best intentions and, second, that systematic processes coupled with human factors principles and practice can be instituted that will at least minimize the problems.

The next chapter provides both background and methods for the successful application of systems and human factors engineering to user-interface development. Particular emphasis is placed on computer aided process control system interfaces.

II

HUMAN FACTORS IN SYSTEMS DEVELOPMENT

Systems Development In Design: Basic Concepts

Systems development (or systems engineering) is the formal process employed in the progression of a man-made system throughout its life cycle. System development originates from the point where a need is identified through system operational testing and continues through decommissioning [16]. It is this rigorous application of procedures, rules, and steps, that take place throughout the creation of a product, that constitute the discipline of systems development. In this context, systems development can also be described as a plan for managing the design process, i.e., a course of action. It is through such a plan that human factors issues and guidelines can be successfully integrated into the total design of a system.

There is no single systematic approach that can be inferred as superior to any other systematic approach. In order to be correct, the systems approach must be individually tailored to meet the specific characteristics of (a) the type and complexity of the system to be built; and (b) the intended end use of the system. It is not so much the mode for practicing systems development that is important. Rather, the importance lies in the implementation of systems development as a philosophy that is incorporated in the design process. Therefore, it is the purpose of this section to outline a generic framework for implementing a top-down systems approach rather than dwelling on the numerous details and debatable points of view on this subject. The sections that follow will acquaint the designer with the fundamentals of this framework. In addition, the role of human factors design activities as they apply to the systems development process are also discussed. A basic understanding of the principles of systems development, and how they relate to human factors issues during the design process, will show the basis for addressing the user interface throughout the various stages of the system life cycle.

The Need for a Systematic Approach: Benefits and Limitations

The need for a systematic approach at some level throughout the design process cannot be overstated.

Failure to observe basic principles of system development can have devastating consequences on the end product. Failure to incorporate comprehensive systems development can result in (a) incomplete analysis of the requirements of the equipment and the total system; (b) the influence of predetermined attitudes of biases in the designer to one or other mode of design solution; and (c) excessive reliance by the engineer on his/her design experience to the detriment of design analysis (e.g., the "shoot-from-the-hip" approach). Without systems development, the design team can quickly lose sight of a product's original functions. In a long-term development process over several years, many hours and resources can be lost in a continuous cycle between defining requirements and redoing the detailed design. It is not unusual in major design efforts to witness different organizations developing the same system components in parallel while being totally oblivious of what the other group is doing.

The end result can be a less-than-optimal system with a sizeable cost overrun. The real loser at the end is the operator who must deal with the poorly developed system when it is finally delivered. While precious time is spent just getting something up and running, the needs of the operator, and the human factors concerns that go with those needs, are rapidly forgotten.

It is not sufficient to brush off the user interface as something to be fixed after implementation. At project completion, new fixes are usually cost prohibitive, as shown by the relationship between cost and design flexibility presented in Figure 4. As the figure illustrates, human factors considerations of the user interface must be addressed at the front end (while design flexibility is high and costs are minimal) as opposed to the back end. Without a systems development

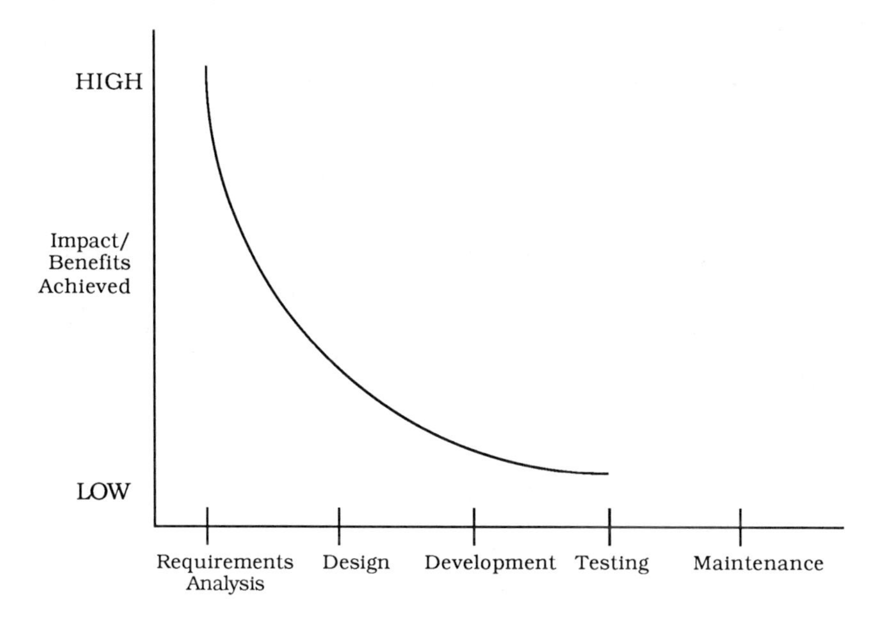

Figure 4. When to apply human factors engineering to systems development (hypothetical relationship).

approach, the human factors specialist falls into the dilemma of being brought into the project at a point where it's really too late to make substantial changes to the system. A project directed under the system development process enables the human factors specialist more opportunity to make significant contributions to the user interface before the design is frozen.

While it can be proven that systems development has demonstrable merit in terms of cost, quality, and final usability, it should not be construed as a cure-all. The implementation of a systems development approach can also be a constraint to successful design. Any design process should be observed in light of the following characteristics identified by Carrol and Rosson [25]:

1. Design is a process; it is not a state and cannot be adequately represented statically.
2. The design process is nonheirarchical; it is neither strictly bottom-up nor strictly top-down.

3. The process is radically transformational; it involves the development of partial and interim solutions that may ultimately play no role in the final design.

4. Design intrinsically involves the discovery of new goals.

It is the responsibility of the designer to recognize the benefits that can be obtained with systems development while being aware at all times of the constraints implied by its implementation.

Some insight to the use of guidelines for the evaluation of existing or design of new information processing systems in the process control industry are presented in the remainder of the book.

Essential guidelines for the evaluation of existing or design of new information processing systems in the process control industry are presented in the following sections.

III

HUMAN FACTORS GUIDELINES AS APPLIED TO VISUAL DISPLAY UNITS

On Standards versus Guidelines

Smith [134] refers to standards as a series of generally stated requirements imposed in some formal way, such as by legislation, by contract, or by management decree. Guidelines, on the other hand, are a series of generally stated recommendations supported with examples, added explanation, and commentary. The criteria and design guidance included in this text are presented as guidelines, not as standards. This text offers guidelines not mandated or endorsed by any legislative authority. The system designer can accept or reject the guidelines, based on personal preference, engineering judgment, or advances in technological knowledge.

The Process of Guidelines Selection

It is important to note that many of the guidelines contained herein are based on (a) sound human factors engineering principles and practice; (b) previous design experience and lessons learned in earlier designs; and (c) extrapolation from experimental study. In the majority of cases, the body of human factors guidelines in the literature is not founded upon rigorous scientific research and inquiry. This is not to say that the guidelines themselves are useless or even deficient. However, it should be recognized by the system designer that a degree of latitude exists regarding the selection of guidelines for any given application. What works for System "A" may not be appropriate for System "B." All guidelines are open to differing degrees of interpretation. In no case is this text an attempt to be the last word on the issues of guidelines for human-computer interaction.

Due to the recent history of guidelines development for computer interaction, it is not unusual to come across guidelines that conflict with one another. The guidelines prepared for this text were studied to resolve such conflicts. In cases where this was not possible, the authors supplied additional commentary for determining when to use one guideline over another. The ultimate tiebreaker for choosing a guideline should rely on the needs of the operator who must use the system when the designers have left. It is never too early (or for that matter too late) to bring the operator into the design loop for consultation and/or operational testing in regard to various guideline options.

Tailoring/Adapting Guidelines for Specific Applications

Generic guidelines, as commonly reported, are of little use to the front-line system engineer or designer, unless they are translated in terms of specific design rules. Design rules, as described by Smith [134], are a series of explicit software design specifications for a particular system application, stated so that they do not require further interpretation by software designers. In short, the system designer or designated human factors engineer on the project has not done his/her job if guidelines, instead of design rules, are part of the system specification. For example, consider the following guideline for messages:

> Messages should be worded concisely, using consistent grammatical structures, phrasing, and punctuation [150].

A design rule based on the above guideline could be

> Alarm messages will contain the instrument ID, the short descriptive name, and the value which is out of tolerance [123].

It has been the experience of Richards [123], as well as the authors, that design rules can be enforced, whereas design guidelines cannot. The software engineer should not be required to interpret the guideline. All the vagaries of guidelines should be handled during the preparation of the design specification.

Prior to the development of the design specification, and for long-term projects where similar systems may be implemented at future dates, it may also be advantageous to consider publishing a document of in-house human factors standards. These standards can be based upon the design rules and guidelines developed for the specific application. With this type of system documentation, the output of effective design criteria can be retained for future reference and not lost when new system design efforts are initiated.

Making Guidelines Effective

The application of guidelines should not be inferred to be a cure-all or a shortcut for accomplishing an effective user interface design. That is, the simple rote adherence to a published set of guidelines is not enough. Smith [134] discusses how good design with guidelines should be observed in light of the following cautionary statements:

- Guidelines cannot take the place of experience or expert design consultants.
- Guidelines are not a substitute or replacement for task analysis.
- The use of guidelines, at least initially, may not save work. Instead, they may actually create more work, the payoff being a more efficient and better user interface when completed.

VDU Description

The human factors guidelines and principles for the user-computer interface (UCI) contained in this handbook tend to emphasize design considerations that most directly relate to the elements of

- display screen,
- keyboard (and/or alternate input device, such as mouse, tracker ball, or joystick), and
- applications software.

The primary points of human factors concern about the man/machine interface reside in those elements and are collectively termed a CRT console.[a] The display screen usually consists of a CRT and functions as the operator's window to the world. It serves as the primary output device for display of information to the operator. The display screen can also serve to provide feedback to the operator that a control function has been initiated. The keyboard is the primary input device for the control, entry, modification, or alteration of data by the operator. The keyboard is probably the most established standard for data input, but its availability and general acceptance should not supersede consideration for alternate input devices (e.g., touch panel, mouse, joystick).

Applications software includes graphics, displays, and functions essential for the operator to communicate with the system. Applications software is what the operator sees while interacting with the plant process. Although software reliability is a necessary component of systems performance, the major software concern in this text is how a particular parameter is displayed to the operator. A large number of issues must be addressed prior to the development of effective applications software. For example, should the format for presentation of a variable set be in the form of a trend plot, bar chart, or display? Does the operator benefit from this information or is it just cluttering the screen? If he or she does benefit from the information, when should it be displayed (e.g., on request, or at all times)?

In conclusion, any device or workstation that incorporates a CRT display screen, computer, software, and a keyboard (and/or other alternate input device) can be addressed by the guidelines presented in the text. An example of a typical CRT console is shown in Figure 5.

[a] The terms VDU (Video Display Unit), VDT (Visual Display Terminal) and CRT console are often encountered in the literature and for our purposes, may be considered equivalent.

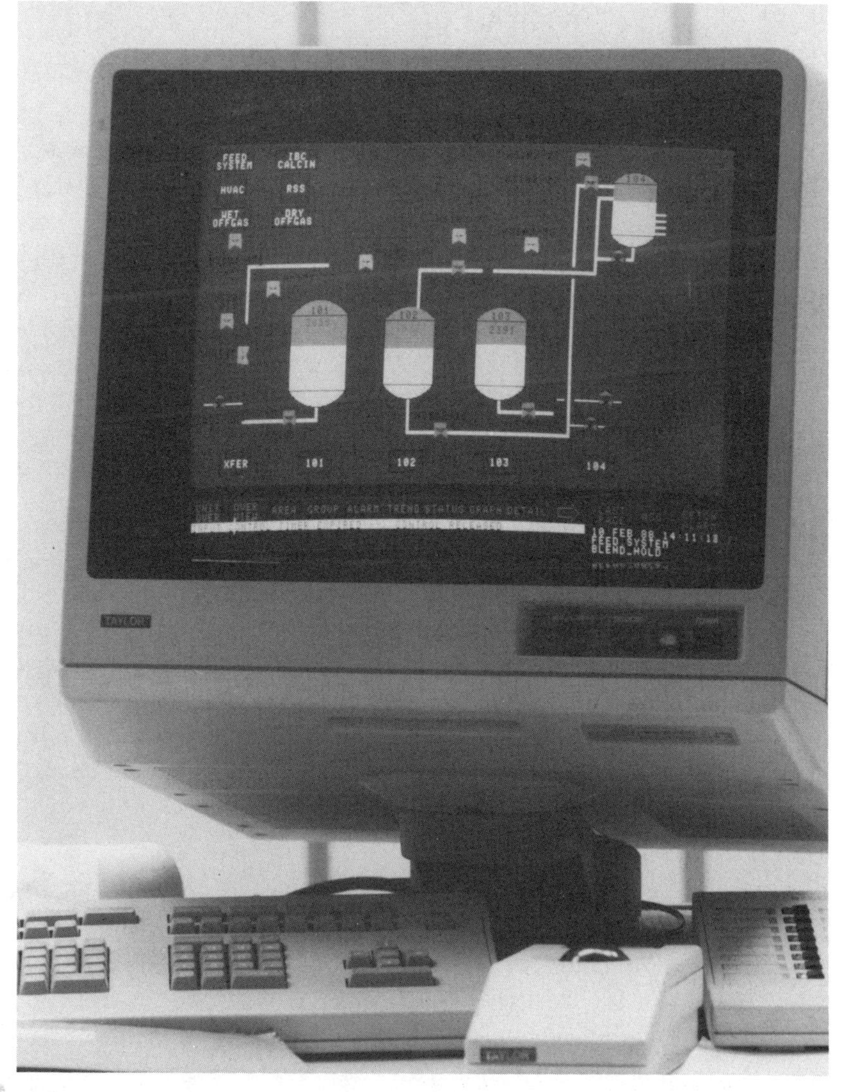

Figure 5. Example of a typical CRT console for process control.

Assumptions

The guidelines contained in this handbook are based on the following several assumptions about VDU equipment and the characteristics of the user population.

VDU Equipment

- Control Room Environment - The environment is assumed to be close to that of a normal office (shirtsleeves environment). The direct analogy between process control rooms and a normal office is, admittedly, somewhat limited (e.g., noise and lighting may be different). However, this assumption precludes the need for addressing special clothing such as anticontamination suits or masks by the operators under routine operations. Those requirements may affect display resolutions, coloration, dimensions and/or selection of input devices.
- Display Medium - The guidelines are based on criteria applicable to a standard CRT monitor. This rationale is based on the general acceptance, versatility, and availability of information regarding the CRT monitor relative to other display technologies (e.g., plasma displays, flat panel, and optical disk.) With minor exception, color and graphics capability are also assumed to be standard features for the CRT monitors. In addition to treating the CRT monitor as the primary display medium, this text presents, for the reader's inspection, a section of guidelines devoted to hardcopy printer characteristics.
- Keyboard Arrangements - The guidelines assume that the keyboard layout for the alphanumeric characters will be of the standard QWERTY arrangement. Trade-off considerations between the QWERTY, Dvorak, alphabetic, Klockenberg, and other alternative configurations will not be discussed. This assumption does not rule out suggestions for altering placement of special function keys, cursor keys, character

keys, or a numeric pad. Those features are not fixed into place or standardized, as is the case for the alphanumeric keys.

User Population

- Physical Characteristics – The user is assumed to have normal (or correctable to normal) visual acuity and color vision. The primary user is also assumed to be an adult without handicaps that require the use of special prosthetic devices. Design for these conditions is achievable but not within the scope of this document.
- Education and Training – The user is *not* assumed to possess special training in computer science or typing skills. Some will have formal training and a high degree of expertise in computer related skills, and others will have almost none. The majority of users, however, will most likely possess military and/or college course work in engineering fundamentals with hands-on experience with process control systems. A second level of users have been identified, but they are much more difficult to define. This group will consist of some combination of process engineers, managers, technicians, and maintenance personnel.
- Task Functions – It is assumed that the majority of users will use a workstation to control and monitor the status of a process control system. The tasks are varied for the identified users, and many of the guidelines are certainly applicable to a multitude of other applications such as military command, control, and communication (C^3).

IV

ORGANIZATION OF GUIDELINES

Overview

The guidelines in this book have been sorted first by major categories and then by variables within each of the subcategories. A variable within the context of this document is any parameter or item which may have an effect on human performance with specific application to user-computer interface (UCI) issues, e.g., brightness, contrast, flicker, etc. The strategy, method, and format for selection and organization of these variables is presented below.

Strategy for Selecting the Guideline Variables

The variables selected for this handbook were initially derived from a thorough review of the available literature. Human engineering guidelines, textbooks, and related materials were reviewed to identify the domain of variables related to design and assessment of the UCI, including those related to the workplace. Any variable which could significantly impact human performance was considered. The resulting list was then subject to internal and external peer review by human factors process control and software engineering design professionals. Recommendations were implemented, and the final list of 99 variables was agreed upon for inclusion into the document.

Method of Presentation

Order and Classification

The guidelines herein are classified within four major categories or functional areas. For each major area, the variables are grouped into specific subcategories. In order to acquaint the user with the organizational logic, the definition for each funtional area is discussed below. A detailed definition of each subcategory is provided in the appropriate section prior to the presentation of the individual variables.

Video Displays – Video display refers to the CRT screen but will include other guidelines related to the presentation or output of information to the operator. Guidelines that address software features, such as screen content, structure, organization, coding, and layout, are presented in this section. Likewise, hardware features that effect visual acuity and readability of the screen are also included (e.g., flicker, contrast, resolution, etc.).

Controls and Input Devices – Controls are the primary device for the input of information to the system. The emphasis for UCI issues will be specific to the layout and arrangement of the keyboard; however, alternate input devices are also discussed.

Control/Display Integration – This section addresses those issues of controls and displays that are not feasible to examine in isolation. For example, user dialogue, data transmission, and feedback to user action are reviewed here.

Workplace Layout and Environmental Factors – The design, configuration, and environment for the CRT workstations are addressed in this section.

Format for Presenting the Variables

The scheme for presenting the variables in the document is outlined below:

- The variables, as noted previously, are organized by major category and subcategory. Each subcategory is defined to familiarize the document user with the topic material. A complete listing of all variables is presented at the beginning of each subcategory section.

- The variable name is introduced on a separate line, followed by a comprehensive definition of the variable.

- All guidelines relevant to the variable (including identification of the original source documents) are presented. In some cases undocumented guidelines that were developed by the authors are referenced as "Authors." Where appropriate, supplementary information is presented in a set of footnotes at the end of the guideline listing.

- A description of the level of research support is presented with each guideline according to the following key:

  Y = Evidence of formal experimentation is presented in the literature to validate the rationale for establishment of the criteria;

  L = Only cursory informal experimentation or observations were conducted and more research is needed; evidence of formal experimentation is extremely scarce; and

  N = No known research is available.

- A *Comment* section follows the guidelines. The purpose of the comment section is to discuss the rationale, limitations, specific applications, and caveats for the guidelines. The comment section provides guidance in drawing overall conclusions from the guidelines.

- An Evaluation and Method for Assessment section for determining whether or not a system under consideration complies with the guidelines presented in this text is included as Chapter 9. Many times compliance of a system may be determined by simple inspection. In other instances, compliance review and acceptance may require (a) the use of special equipment, (b) access to the equipment procurement specification, or (c) that the system be operated to determine its real ease of use.

V

VIDEO DISPLAYS

Hardware Aspects

This section discusses variables of interest, definitions, guidelines, notes, and comments pertaining to many of the various hardware aspects of VDUs. The following variables most directly affect display legibility, readability, and discrimination:

- Flicker
- Contrast ratio
- Display luminance
- Phosphor
- Glare
- Screen resolution

Flicker

Flicker is the perception of rapid fluctuations in luminance level or position. It is characterized by an impression of uneven changes or jerky movements. As soon as a character is projected onto the CRT screen, it begins to fade at a rate depending on the persistence of the screen phosphor. To maintain visibility of the character on the screen, the signal must be continually regenerated or refreshed. If the signal is not refreshed often enough, the display will appear to blink or flicker. If the regeneration frequency is adequate, the user will perceive the image as being steady or fused (critical fusion frequency).

This is due, in part, to the psychophysiological phenomenon of persistence of vision.

Factors affecting the severity of flicker are intensity, visual angle, color, age of the user, ambient illumination, and phosphor characteristics.

Guidelines	**Research Support**	**Source**
1. The regeneration rate for a particular CRT display should be above the critical frequency of fusion so that the occurrence of disturbing flicker is not perceptible.[a]	Y	150

[a] In order to produce a flicker-free display, ideal regeneration rates to 50-60 Hz have been recommended, with 60 Hz as the preferred frequency [5], [42], [23]. Displays made with a refresh rate of under 20 Hz are usually very annoying to the viewer, and they are not flicker-free under about 50 Hz [23].

Comment. It can be inferred that a minimum refresh rate of 60 Hz is the most acceptable criterion. However, the objective of these recommendations is to provide a flicker-free display. It would be more appropriate to state a functional requirement, namely (as in Reference [150]), that the display should not appear to flicker under certain specified settings and ambient light conditions.

Extensive research has demonstrated degraded legibility of alphanumeric characters and graphic displays under flicker conditions. There are also negative physiological results from flicker that range from eye strain and headaches to induced epileptic seizures.

Contrast Ratio

Contrast is the difference of luminance, or reflectance, between a figure and its background. The ratio between those two luminances or reflectances is called the contrast ratio. In the case of alphanumeric characters or graphic displays, the characters or lines are considered to be the figure. The background may be either dark or light, depending on whether positive or negative image polarity is used (see also Image Reversal Coding). The contrast ratio is usually reported with the larger number first, whether that number is the luminance of the figure or the background. Note that the contrast ratio numbers must be formed using common luminance units.

Even though the screen is self-luminescent, the contrast ratio of a CRT display screen is affected by ambient lighting conditions (darkened room or a brightly lit room). That is, ambient lighting will affect both the adaptation levels of the user's eyes and the reflected light even from nonglare screens.

Guidelines	Research Support	Source
1. 3:1 minimum; 5:1 to 10:1 preferred; 15:1 maximum.[a]	Y	146, 23, 22 41, 58
2. Background levels 15 to 20 cd/m^2.	Y	23, 22

[a] Reference [41] states a preferred range of 6:1 to 10:1. Reference [58] recommends a 4:1 minimum in an ambient lighting environment of 75 to 100 fc. Reference [23] recommends 8:1 to 10:1 as the optimum. Reference [146] recommends a 10:1 minimum when using white characters on a black background and a 5:1 minimum when using black figures on a white background. Reference [22] recommends a background level of at least 20 cd/m^2.

Comment. CRT displays are used in a broad range of lighting environments. Thus, to suit the ambient environment and user preferences, it seems most reasonable that the luminance levels and contrast ratios of the displays should be adjustable over the entire range listed in Guideline 1. We should note here that the actual luminance level at the retina of the eye is affected by physiological variables and by environmental characteristics such as the use of glasses, contact lenses, antiglare filters, etc. The measured contrast ratio should be evaluated with antiglare filters, etc., in place. Contrast ratios must be adjusted for the use of color displays because of differential sensitivity of the eye to various colors.

Display Luminance

Display luminance refers to the total light from a display. This light output may be either emitted light or reflected light. In the case of CRT displays, the light is emitted. The optimum luminance for CRT displays is partially determined by contrast ratios, ambient light levels, and operator variables. The display luminance can be calculated on the basis of the brightest portion of the display or on the basis of the integrated luminance over the entire area of the display.

	Guidelines	**Research Support**	**Source**
1.	45 cd/m^2 minimum, with 80 to 160 cd/m^2 preferred.[a]	Y	23, 58, 69
2.	10 cd/m^2 minimum average level.	Y	74

[a] Guideline 1 refers to the integrated level over the whole display screen when filled with text. The luminance level that will provide good acuity partially depends on the level of other work that must be referred to during the task (such as manuscripts, etc.). Too great a mismatch between the display screen and documents causes the operator to have to shift both focus and adaptation level–although this effect is not so great as had been previously thought [139]. Reference [58] recommends a minimum display luminance of 85 cd/m^2.

Comment. Other factors in the surroundings and in the operator (fatigue, acuity, time of day, etc.) also affect the optimum luminance level. The simple solution and the one recommended by these authors is probably to provide the operator with a means of adjusting the luminance level of the display. The range of adjustment should encompass the high and low values stated in Guideline 1.

Phosphor

Phosphor is a substance which emits light when excited by radiation, and the degree to which it continues to emit light after radiation has ceased is referred to as its persistence. The phosphor specification which follows is applicable only to monochrome CRT displays.

Guidelines	Research Support	Source
1. A green phosphor should be used.	Y	23
2. A medium persistence phosphor should be used.	Y	45

Comment. A green phosphor is recommended since the human eye is most sensitive to light in this visible spectrum. However, there is no empirical evidence to indicate that alternative colors (orange, white, yellow) would be a hindrance to operator performance [23].

Color choice among phosphors is secondary to the factors of display clarity and contrast in determining test legibility. Therefore, color is more a matter of personnel preference. In order to avoid blurring or smearing of images, medium-persistence phosphor is recommended.

Glare

Glare is the presence of areas of high luminance reflectance on the display. Glare reduces contrast ratios and adversely affects pupil aperture size and focus.

Guidelines	Research Support	Source
1. The screen should be positioned so that sources of light and/or bright objects do not reflect into the expected viewing position.	Y	45
2. The surface of the VDU screen should be modified to reduce specular glare.[a]	Y	45

[a] Specular glare sources may also be transient. A person wearing a white shirt and standing behind the VDU user will often create glare on the screen.

Comment. Film coatings, roughened surfaces, and mesh filters appear to accomplish glare reduction equally well. In each case, the surface treatment may degrade the quality of the displayed elements but, for the occasional user, the reduction in glare far outweighs the slight fuzziness of the display.

Caution should be used in the selection of gauge glare-filter screens, however, because some of these screens reduce luminance and contrast ratios below the acceptable minimums.

Hoods are sometimes attached to the display to reduce glare but this is not recommended [45]. Hoods are cumbersome and tend to restrict the viewing angle of the screen, whereas the screen treatments are inexpensive and can be added to any existing equipment.

Screen Resolution

Screen resolution determines the fineness of detail that can be displayed.

Guidelines	Research Support	Source
1. Regardless of whether the display is raster scanned or directly addressed, it should maintain the illusion of a continuous image. The viewer should not have to resolve scan lines or matrix spots.	L	142

Comment. Screen resolution is totally dependent on what displays are desired. That is, screen resolution should be sufficient for the level of detail required for the task at hand. Commercially available television monitors are sufficient for the purpose of reading text and coarse drawings.

Screen Structures and Content

Screen structures and content include all elements or "primitives" that comprise a screen.

- Cursor
- Text
- Labels
- Messages
- Abbreviations
- Error statements
- Nontextual messages
- Data display
- Data entry
- Instructions

Specific applications for organizing, grouping, and layout of the screen are discussed in the section on screen organization and layout. Special coding and highlighting techniques have also been reserved for a later section.

Cursor

A cursor is a dot, block, or special marker. The cursor can be used to indicate the position on the screen where the operator's attention should be focused. It also serves as the carriage position on a typewriter. Continuing with the typewriter analogy, the cursor indicates the position in the display where the next character will appear. There are several general guidelines related to cursor design and its characteristics.

	Guidelines	Research Support	Source
1.	The cursor should be easily seen but should not obscure the reading of the character or symbol it marks.	Y	41, 23
2.	The cursor should be easy to move from one position to another.	Y	42, 23
3.	The cursor when used on a monitoring task, should blink at about 3 Hz.	Y	42, 136
4.	The cursor should not be so distracting as to impair the searching of the display for information unrelated to the cursor.	Y	42, 23

Comment. In conjunction with the guidelines listed above, Cakir et al. [23] generally believe the cursor (for most applications) should be in the form of a box or block with a 3-Hz blink, rate to be preferred. The underscore cursor is not recommended if underscoring is used to serve another function on the display or when scanning is used to locate the cursor.

Text

This rather broad variable includes the multitude of factors that affect the legibility and comprehension of textual information. From the perspective of screen design, perhaps the two most relevant factors are those associated with textual format and content. In short, textual format describes essentially how the text should best be presented, and textual content entails what should be displayed. Basic structural features are other primary characteristics to be addressed. In this section, the emphasis of text characteristics is limited to prose. Other variables, such as message and labels, will be discussed separately.

Guidelines	Research Support	Source
1. Consistent format should be maintained from one display to another.[a]	Y	135
2. Prose should be displayed conventionally, in mixed upper and lowercase.[b]	Y	135
3. Displayed paragraphs should be separated by at least one blank line.	Y	135
4. In textual display, every sentence should end with a period.	Y	135
5. In textual display, short, simple, concise sentences should be used.	Y	135

[a] Careful consideration of form and sequencing are more important the less prior knowledge one expects the reader to have of the subject [135].

[b] Uppercase should be used when lowercase letters will have decreased legibility, which is true on display terminals that cannot show descenders for lowercase letters [135].

Comment. The interplay of such diverse factors as character size, contrast, image stability, and character spacing impact the overall legibility of textual information. In many cases, the interplay of those factors make it extremely difficult to develop guidelines for optimizing text legibility and comprehension. Preliminary research indicates that there is no comprehension difference between text presented on a CRT screen when compared with printed pages [103]. Other findings by Gould, et al. [61], however, indicate that reading text from the CRT is slower than reading from the printed page.

Other research has been devoted to the development of readability formulas or mathematical equations consisting of various properties considered in text legibility (e.g., word length, sentence structure). The utility of those formulas as a valid method is still a matter of much debate.

In conclusion, with the exception of text structure, specific guidelines are limited for textual form and content. Such guidelines that do exist are entirely too general to be of much value. In addition to the high interplay of various factors, the unique task-dependent nature of textual comprehension on a CRT screen may also hinder the development of suitable guidelines.

Labels

Proper labeling is the unambiguous designation of structures and functions presented on VDU displays. It also refers to the act of placing a descriptive title, phrase, or word adjacent to a group of related objects or information [47]. The Eastman Kodak edition of "Ergonomic design for People at Work"[42] defines labels, (and signs as well) as "short messages used to transfer information about policies or equipment use between people." That document also identifies three considerations in label design: (a) comprehensibility, a measure of how reliably the receiver interprets the information; (b) legibility, the user's ability to discriminate among or recognize letters or numbers; and (c) readability, the ease of reading words or labels, assuming that the requirements for specific legibility are satisfied.

There is considerable overlap in the guidelines between labels, and associated variables such as text and messages. Therefore, the reader is referred to these relevant variables for additional information as needed. In general, perhaps the most relevant factor for making a distinction between text, messages, and labels, can be derived by applying the following rule of thumb based on word length. Text can be defined as any prose of three lines or longer. Labels are usually confined to less than one line. A Message is any string of words falling between the criteria for text and label.

	Guidelines	Research Support	Source
1.	Labels should convey the basic information needed for proper identification, utilization, actuation, or manipulation of the item.	Y	146
2.	Labels should be consistent with such factors as a. Accuracy of identification required, b. Time available for recognition or other responses, c. Distance at which the labels must be read, d. Illuminance level and color, e. Criticality of the function labeled, f. Vibration/motion environment of the user.	Y	146
3.	Labels should be horizontal and read from left to right.[a]	Y	146
4.	Labels should be placed on or very near the items which they identify.	Y	146
5.	For single data fields, the label should be placed to the left of the data field.	N	5
6.	For repeating data fields, the label should be placed above the data fields.	N	53
7.	Labels should primarily describe the functions of items.[b]	Y	146

8.	Control labeling should indicate the functional result of control movement (e.g., increase, ON, OFF).	Y	146
9.	Control and display labels, should convey verbal meaning in the most direct manner by using simple words and phrases. Abbreviations may be used when they are familiar to operators (e.g., psi, km).	Y	146
10.	Words should be chosen on the basis of operator familiarity whenever possible, provided the words express exactly what is intended.[c]	Y	146
11.	Similar names for different controls and displays should be avoided.	Y	146
12.	The units of measurement (e.g., volts, psi, meters) should be labeled on the screen or panel.	Y	146
13.	Labels should be printed in all capitals; periods should not be used after abbreviations.	Y	146
14.	When dealing with mechanical labeling, to reduce confusion and operator search time, labels should be graduated in size.[d]	Y	146
15.	Label names should be easily discriminated from surrounding labeled fields or messages.	Y	47
16.	Labels for data fields should be distinctively worded or highlighted so that they will not be readily confused with data entries, labeled control options, guidance messages, or other displayed material.	L	5
17.	Where entry fields are distributed across a display, a consistent format should be adopted for relating labels to entry areas.	Y	5
18.	Where a dimensional unit (gpm, cm, deg, etc.) is consistently associated with a particular data field, it should be part of the fixed label, not entered by the user.	L	5

[a] Labels and information thereon should be oriented horizontally so that they may be read quickly and easily from left to right. Vertical orientation may be used only when labels are not critical for personnel safety or performance and where space is limited. When used, vertical labels should read from top to bottom [146].

[b] Engineering characteristics or nomenclature may be described as a secondary consideration [146].

[c] Brevity should not be stressed if the results will be unfamiliar to operating personnel. For particular users (e.g., maintenance technicians), common technical terms may be used even though they may be unfamiliar to nonusers. Abstract symbols (e.g., squares and Greek letters) should be used only when they have an accepted meaning to all intended readers. Common, meaningful symbols (e.g., % and +) may be used as necessary [146].

[d] The characters in group labels, should be larger than those used to identify individual controls and displays. The characters identifying controls and displays should be larger than the characters identifying control positions. With the smallest characters determined by viewing conditions, the dimensions of each character should be at least 25% larger than those of the next smaller label [146].

Comment. The majority of these guidelines are extracted from the standards established in MIL-STD-1472C [146] and NUREG-0700 [150]. Their origins are founded in hardware labels, as opposed to CRT labels. However, they should still be applicable. The guidelines from References [5], [47], and [53] are more solidly founded in specific CRT applications. In spite of some overlap, they comprise a relatively comprehensive set of guidelines for labeling.

Messages

A computer system communicates with users through messages. A message may be a prompt, a diagnostic message generated by an error condition, or an information or status message. By applying the rule of thumb discriminating between text, labels, and messages, any word string greater than one line but less than three can be classified as a message. Thus, a message is more detailed than a label and tends to provide description or direction. In comparison, a label is a simple identifier of some object or entity. Message length and load are two concepts frequently addressed when message comprehension issues are examined. Huchingson [67] defines length as "the number of words, or characters, that are displayed at once on one or more lines of a sign or sequence." Load is described as "the number of units of information displayed (within a message)." Units are the basic elements or building blocks which make up the information given. Each unit provides an answer to a question (e.g., what, where, who). Manipulation of those factors drives the message design. The reader is referred to the sections on text and labels, and the subcategory Alphanumeric Characters for additional information relevant to messages.

Guidelines	Research Support	Source
1. The computer should be capable of providing two levels of detail.[a]	N	5
2. Messages should be strictly factual and informative.[b]	N	47, 53
3. Message dialogue should not be hostile to the user.	N	5
4. Messages should be constructed using short, meaningful, and common words.	L	53
5. The message should consider the prior knowledge of the user and the user's context.	L	90
6. Sentences should be kept as simple in structure as possible.	L	67
7. Messages should require no transformations, computing, interpolation, or reference searching.	L	47
8. Messages should be stated in the affirmative and preferably in the active voice.	Y	5
9. Items to be remembered by the user should be placed at the beginning of the message.	Y	101
10. Items to be recalled by the user should be placed at the end of the message.	Y	101

11.	Items of lesser importance should be placed in the middle of the message.	Y	146

[a] The "rule of two" suggests that the computer have the capability of providing two levels of detail: level 1–directed to inexperienced user containing detailed messages and level 2–directed to experienced users and containing abbreviated messages [5].

[b] The tendency in the past few years has been to avoid emotional implication and "cute" messages, as they tend to distract or alienate rather than inform.

Comment. The key point for these guidelines revolves around the issue of user comprehension. In other words, is the message under evaluation immediately understood by the user? Does the user understand what the message designer meant? A clarification of these basic questions with the potential users should ensure the overall usefulness and success of the message set.

Abbreviations

Abbreviating is the process of reducing words or phrases into a shortened form while retaining the original meaning. This shortened form of communication is a highly effective method for presenting a concept where space is limited and/or speed of recognition is essential.

Abbreviations should be used if they are significantly shorter, save needed space, and will be understood by the prospective users.

The challenge to designing effective abbreviations resides in the ability to shorten words/phrases while still retaining their intended meaning to the users.

	Guidelines	**Research Support**	**Source**
1.	Only standard and commonly accepted abbreviations should be used.	Y	150, 146
2.	Abbreviations should be short, meaningful, and distinct.[a]	L	135, 18

3.	The system should permit abbreviations of commands.[b]	L	18
4.	Whenever possible, experienced users should be provided with a set of abbreviations for frequently used commands.	Y	5
5.	Abbreviations should be consistent in form.	Y	135
6.	A dictionary of abbreviations should be available for on-line user reference.	Y	135
7.	Abbreviations and acronyms should not include punctuation.[c]	Y	135

[a] Resist the temptation to use abbreviations the user is not likely to understand or remember just to make room for more data on the display [18].

[b] The system should recognize a command from the first, second, or third letters (as necessary to make it distinct from other commands). Novice users can type in the entire command, while experienced users can abbreviate or truncate it. The system should accept either form of the command [18].

[c] For example, the acronym for cathode ray tube should be "CRT" and not "C.R.T [35]."

Comment. Research points out that rule-based abbreviations are more easily comprehended when compared with ones that are not rule-based. Bailey [5] has referenced a general set of rules for abbreviating single words and word groups and a more specific set of rules for abbreviating commands. As an aid to the user of this document, those sets of rules are summarized below. Note that the method of creating abbreviations differs between the two sets of rules.

When abbreviating a single word,

- Determine the number of characters required in the abbreviated term.
- Remove suffixes such as -ed, -es, -er, and -ing from the word to be abbreviated.

- Choose the first letter and last consonant of the word to be abbreviated as the first and last letter of the abbreviation. For example, Boulevard would be B--D.
- Fill the remaining spaces with the consonant in the order in which they appear in the word. Avoid the use of double consonant in the abbreviation. For example, a five-character code for boulevard would be BLVRD, a four-character code would be BLVD, and a three-character code would be BLD.
- If there are insufficient consonants, use the first vowel in the order in which it appears in the word. For example, a six-character code for boulevard would be BOLVRD.

When abbreviating two-word groups,

- Determine the number of characters required in the abbreviated term.
- Take half of the characters from the first word and the other half from the second word, using the above abbreviation methodology for single words. If an odd number of total characters is needed, take the odd character from the longer word. For example, a seven-character code for PROGRAM TEST would be PRGM TST.

When abbreviating commands,

- For terms consisting of more than one word, create an acronym by taking the first letter of each word.

When abbreviating monosyllabic words,

- Take the initial letter of the word and all subsequent consonants.
- Make double letters single.
- If more than four letters remain, retain the fifth letter if it is part of a functional cluster (such as th, ch, sh, ph, or ng); otherwise, truncate from the right. Delete the fourth letter if it is silent in the word.

Bailey notes that exceptions should be made only when special significance is achieved in selecting one letter or group of letters over another. These rules are meant only for the development of new abbreviations and should not be applied to existing standard abbreviations.

Two general observations can be drawn from the guidelines: (a) ensure that abbreviations are understood by and meaningful to the user, and (b) verify that the abbreviations are standardized and consistent.

Error Statements

The computer system's major mode of communication with the user is accomplished through various messages which are presented on the screen (e.g., normal prompts, advisory messages, system response to commands). The development of useful and meaningful error statements is generally considered critical to user interaction. This section emphasizes that special class of messages.

Error messages are intended to guide new, inexperienced, and (occasionally) expert users in the efficient use of a system. Clear, concise, and meaningful error messages will increase user acceptance and user ratings of a display system as well as the speed with which the process is completed [18].

General principles of message design have been examined in a previous section, and the reader is referred to that section for supplemental guidelines. The issue of errors in UCI guidelines is enormous, and error statements are only one component of this greater domain. This text omits the research associated with error guidance, correction, and control of error and focuses instead on the specific structures for presenting error statements to the user.

Guidelines	Research Support	Source
1. The computer system should contain prompting and structuring features by which an operator can request corrected information when an error is detected.	L	150

2.	Error messages should be worded as specifically as possible. When the computer detects an entry error, an error message should be displayed to the user stating what is wrong and what can be done about it; e.g., "No left parenthesis used."	L	135, 18
3.	The wording of error messages should be appropriate to a user's task and level of knowledge.	L	135, 18
4.	When a data entry or (more often) a control entry must be made from a small set of alternatives, those correct alternatives should be indicated in the error message displayed in response to a wrong entry.	L	135
5.	Error messages should be stated in polite but neutral wording, without implications of blame to the user, without personalization of the computer, and without attempts at humor.	L	135, 18
6.	Following the output of simple error messages, the user should have the option of requesting more detailed explanation for errors (e.g., successively deeper levels of explanation should be provided in response to repeated user requests for HELP).	L	135, 18
7.	When multiple errors are detected in a combined user entry, some indication should be given to the user, even though complete messages for all errors cannot be displayed together.	L	135
8.	Error messages should be output two seconds after a user's entry has been completed.[a]	L	135
9.	System documentation should include as a supplement to on-line guidance, a listing and explanation of all error messages.[b]	L	135, 18
10.	Following error detection, users should be prompted to reenter only the portion of a data/command entry that is not correct.	L	150, 135, 18

11.	In addition to a clear text error message, an error identification number (ID) should precede each message.[c]	L	18
12.	Error messages should always state or clearly imply at least a minimum of (a) what error has been detected and (b) what corrective action to take.	L	18
13.	If an error is detected in a group of stacked entries, the system should process correct commands until the error is displayed. A suitable error message should be presented, and no more inputs should be processed until the error is corrected.[d]	L	18

---

[a] In general, the display of error messages should be timed so as to minimize disruption of the user's thought process and task performance [135].

[b] Documentation of error messages will facilitate review of that aspect of user interface design, since it is difficult to generate all possible error messages by actually making errors in on-line transactions [135].

[c] An error identification number permits the experienced user to recall the kinds of procedural errors that might have been made; it also allows the novice user to refer to system documentation for more details on the possible reasons for the message, the place the error was made, and the possible remedies. Error messages are distinct from edit prompts which refer to format or spelling errors. The form of the error ID (usually only one error message at a time) is Error nnn. (message). A space of three digits should be provided for the ID. Leading zeros or spaces should not be required in referring to an error message [18].

[d] Allowance for the capabilities of skilled users is important. System design should allow skilled users to make inputs faster than they can be processed and displayed by the system. This can be done by allowing the user to make entries to screen options which are not yet displayed by stacking entries (i.e., answering several screens ahead) [18].

Comment. The basis of these guidelines is a reflection of the general principles for message design; this is to be expected, since errors are indeed a type of message. Similar observations are echoed throughout both variables: make the error message meaningful, understandable, brief, and comprehensible to the user. The guidelines tend to circle around this central issue with an assortment of techniques for satisfying those requirements.

Nontextual Messages

Text refers to alphabetic character strings arranged to form a meaningful word or words, and nontextual information is commonly referred to as alphanumeric coding. Alphanumeric strings of a nontextual nature are arranged to form an identifiable code. This includes natural codes taken from common usage and arbitrary codes which usually require learning.

Further classifications of these alphanumeric codes can also be associative or transformational. Associative coding serves as a stimulus with some unique response. Telephone numbers generally fall into this category. Transformational coding applies a strict set of rules to the data in order to derive a code. For example, the code CV-P004-091 designates the system, location, and type of control valve in a nuclear power plant. This section focuses on coding principles and nontextual data which involve the use of alphanumeric characters. Special types of guidelines for other types of coding, such as color, brightness, and blinking, are addressed in later sections.

	Guidelines	**Research Support**	**Source**
1	When using alphanumeric codes, a consistent convention should be adopted that all letters shall be either uppercase or lowercase	L	135
2.	When codes combine letters and numbers, characters of each type should be grouped together rather than interspersed.	L	42, 53, 18
3.	Meaningful codes should be adopted in preference to arbitrary codes; e.g., a three–letter mnemonic code (DIR = directory) is easier to remember than a three-digit numeric code.	L	53, 40
4.	When arbitrary codes must be remembered by the user, they should be no longer than four to five characters.	L	42, 135, 53, 18
5.	Code length and format should be constant throughout any single category.	Y	53

6.	Codes should contain predictable letter sequences.	Y	53
7.	Long codes (seven or more characters) should be broken into three- or four-character groups; i.e., separate groups by a hyphen or blank space.	Y	42, 47, 53
8.	Refrain from using 1's and 0's in code vocabularies.	Y	53

Comment. Codes using uppercase labels, will be somewhat more legible than those using smaller case labels. For data entry, computer logic should not distinguish between upper and lowercase codes, because the user will find it hard to remember any such distinctions [135]. Codes based on common English usage are the most easily used because they require minimum learning. Arbitrary codes are seldom easy to use [53].

Reference [53] recommends a maximum of six code characters. However, that reference also cautions that as the length of a field increases, errors in its use will also increase. Reference [42] recommends that an all-digit code should be used where possible.

Codes containing predictable letter sequences can be keyed more rapidly. For example, the letter combinations "TH" and "IN" are much more predictable than "YX" or "JS," and can be keyed faster [53].

Reference [53] notes that about half of all coding errors could be eliminated if 0 and 1 were not in the alphanumeric code vocabulary. In addition about two thirds of all errors could be eliminated if 1, 0, 8, and B were not used in codes. Reference [42] recommends that the letters B, D, I, O, Q, and Z and the numbers 0, 1, and 8 should be avoided.

The main emphasis for these guidelines centers around suggestions and techniques for minimizing errors in data entry and comprehension (e.g., errors of omission, addition, substitution, and transposition). It is extremely difficult to change a coding system once the code has become part of the production process. Therefore, caution should be exercised when making recommendations for altering an alphanumeric (as well as all modes of coding) scheme after it is in place. Many of the guideline sources are in agreement with each other with one exception pertaining to the mixing of digits and alphabetic characters. Galitz [53] and Brown et al. [18] suggest character types and numerals should be grouped together (e.g., HW5 rather than H5W). In contrast, Huchingson [67] pointed out that, for three-digit numeric codes, a letter between two numbers would minimize the danger of transposition errors. For example, 3H4 is more accurately recalled than 34H. More research is needed before this conflict can be resolved.

Data Display

Data display refers to the various techniques and approaches for the presentation of data so that the operator can obtain the needed information in a timely and accurate manner. In addition to timeliness and accuracy, efficient information assimilation and minimal memory load on the operator are other desirable features of effective data display. Data can be presented via a CRT monitor or a hardcopy printer. Data, as it applies to the guidelines in this section, refers to the fundamental building blocks or raw materials the operator needs to obtain the needed information. The terms information and data are often used interchangeably, but basic differences exist in their meanings. As previously described, data are merely the building blocks of information, whereas information may be regarded as the answer to a question based on the extraction of the raw materials comprising the data. Data display differs from text display in content and structure. The guidelines for text display generally assume a prose style comprised of alphabetic characters. In data display, the arrangement and order of numbers (and coded messages) are emphasized.

Guidelines	Research Support	Source
1. Displayed data should be tailored to user needs, providing only necessary and immediately usable information at any step in a transaction sequence.	L	135, 18
2. Data should be displayed to the user in directly usable form.[a]	L	135, 53

3.	Data should be consistent, following standards and conventions familiar to the user.		135
4.	When protection of displayed data is essential, do <u>not</u> permit a user to change controlled items.	L	135
5.	In general, do not require the user to rely on memory, but recapitulate needed items on the succeeding display.	L	135
6.	The detailed internal format of frequently used data fields should be consistent from one display to another.	L	135
7.	Long data items of arbitrary alphanumeric characters should be displayed in groups of three or four separated by a blank.[b]	L	135, 53
8.	In tabular displays, columns and rows should be labeled following the same guidelines proposed for labeling the fields of data forms.	L	135, 18
9.	In tabular displays, the units of displayed data should be consistently included in the column labels, or following the first row of entry.	L	135
10.	Columns of numeric data without decimals should be displayed right-justified; numeric data with decimals should be justified with respect to the decimal point.[c]	L	135, 53, 18
11.	Lists of alphabetic data should be vertically aligned with left–justification to permit rapid scanning; indentation can be used to indicate subordinate elements in hierarchic lists.[d]	L	135, 53
12.	Data lists should be organized in some recognizable order, whenever feasible, to facilitate scanning and assimilation; e.g., dates may be ordered chronologically, names alphabetically.	L	135, 53
13.	Listed data should be distinctive from lists of menu options.	L	135
14.	When listed items are labeled by number, the numbering should start with 1 and not 0.	L	135

15. For hierarchic lists with compound numbers, the complete numbers should be used, rather than omitting the repeated elements; i.e., — L — 135

    2.1 Position Designation
        2.1.1 Arbitrary Positions
            2.1.1.1 Discrete
            2.1.1.2 Continuous

16. In dense tables with many rows, a blank line (or some other distinctive feature) should be inserted after every fifth row as an aid for horizontal scanning. (If space permits, a blank line after every third row is even better). — L — 135, 53

17. When data are displayed in more than one column, the columns should be separated by at least three to four spaces if right-justified and by at least five spaces otherwise. — L — 135

18. When tables are used for referencing purposes such as an index, the indexed material should be displayed in the left column, the material most relevant for user response in the next adjacent column, and associated but less significant material in columns further to the right. — L — 135

19. Longer series of strings or lists of data should be organized in columns to provide better legibility and faster scanning. — L — 18

20. If data are to be entered from paper forms, the design of the input screen and the layout of the paper form should correspond. This helps the user to find and keep a location while looking back and forth from the form to the terminal. — L — 18

21. Each list of selections should have a heading that reflects the question for which an answer is sought; e.g., — L — 18

Select Plant Mode

1. Start up
2. Steady-state operation
3. Shutdown

22.	In a list of options, the most frequently used options should be placed at the top of the list.[e]	L	18
23.	Selection numbers should be separated from text descriptors by at least one space. Include space after the period, if used. Right-justify selection numbers.	L	18
24.	When lists or data tables extend beyond one display page, the user should be informed when a list is or is not complete.[f]	L	18
25.	Labels for single data fields should be located to the left of the data field and separated from the data field by a unique symbol (such as a colon) and at least one space,[g] e.g., TITLE: ____________________	L	53
26.	When caption sizes are relatively equal, both captions and data fields should be justified left. One space should be left between the longest caption and the data field column; e.g., FEED FLOW: __________________ PRIMARY COOLANT: _____________	L	53
27.	When caption sizes vary greatly, captions should be right-justified and the data fields should be left-justified. One space should be left between each caption and the data field;[h] e.g., FEED FLOW: _________ STEAM GENERATOR PRESSURE: _________	L	53
28.	A field group heading should be centered above the captions to which it applies. It should be completely spelled out and related to the captions; e.g.,	L	53

- - - - - - - - - - - - - DRYWELL - - - - - - - - - - - - -

LEVEL (FT.)	PRESSURE (PSIG)	TEMPERATURE (F)
________	________	________
________	________	________

29. When section headings are located on the line above related screen fields, the captions should be indented a minimum of five spaces from the start of the heading;[i] e.g., L 53

SECONDARY CONTAINMENT
     LEVEL: ________
     PRESS: ________

30. When section headings are placed adjacent to the related fields, they should be located to the left of the topmost row of related fields. The column of captions should be separated from the longest heading by a minimum of three blank spaces;[i] e.g., L 53

SECONDARY CONTAINMENT   LEVEL: ____________
                          PRESS: ____________

31. At least five spaces should appear between the longest data field in one column and the rightmost caption in an adjacent column;[j] e.g., L 53

ALARM: __________   DATE: __________
STATUS: ________   TIME: __________

32. Where space constraints exist, vertical lines may be substituted for spaces for separation of columns of fields. L 53

33. For multiple-occurrence fields without group headings, at least three spaces should exist between the columns of fields;[k] e.g., L 53

PWR	LVL	TEMP
________	________	________
________	________	________

34. For multiple-occurrence fields with group headings, at least three spaces should appear between columns of related fields and at least five spaces should appear between groupings;[k] e.g., L 53

- - - - - - UNIT I - - - - - -			- - - - - UNIT II - - - - - -		
PWR	LVL	TEMP	PWR	LVL	EMP
_____	_____	_____	_____	_____	_____
_____	_____	_____	_____	_____	_____

[a] The user should not be required to transpose, compute, interpolate, translate displayed data into other units, or refer to system documentation to determine the meaning of displayed data [135].

[b] As an exception to Guideline 7, words should be displayed intact, whatever their length. Also, grouping should follow convention where a common usage has been established, as in the NNN-NN-NNNN of social security numbers [135].

[c] Do not add or remove zeros arbitrarily after a decimal, since in some applications the zeros may affect the meaning in terms of significant figures [18]. In the interests of compact display format, a short list (of just four to five items) may be displayed horizontally on a single line if done consistently [135].

[d] In the interest of compact display format, a short list (of just four to five items) may be displayed horizontally on a single line if done consistently [135].

[e] Reference [53] recommends that for lists of up to seven alternatives, the most probable alternative be placed at the top. In longer lists, or in short lists when there is no obvious frequency or pattern, the items should be placed in alphabetical order.

[f] Lists or data tables often extend beyond the amount that can be shown on one display page. The user must be informed when a list is or is not complete. Unless the list is short and it is obvious that it does not fill the available space, it should be marked with the message – END OF LIST– Incomplete lists should be marked–CONTINUED ON NEXT PAGE – If the message is too long and it would reduce the display capacity, it may be shortened to -- CONTINUED -- [18].

[g] Caption rules are most similar to those for data entry screens without source documents. For multiple-occurrence fields, the caption will be centered above its related data field, and the data field need not be preceded by a unique symbol such as a colon. Presence of a field will be communicated by the display of data itself. If the field contains no data, it is irrelevant to the user. The overriding caption consideration is that it be clear and easily identifiable as a caption [53].

[h] Left-justification of captions and right-justification of data fields will usually yield a more balanced display [53].

[i] Scanning an inquiry screen will be aided if logical groupings of fields are identified by headings. This permits scanning of headings until the correct one is located, at which point the visual search steps down one level to the items within the grouping itself. Guidelines 29 and 30 are intended to provide easily scanned headings [53].

[j] The design goal is visual separation of columns of fields. Note that these separation guidelines are minimums. If wider spacing is feasible, utilize it [53].

[k] Multiple-occurrence fields must also provide adequate visual separation of columns of fields [53].

Comment. In spite of the relative impact of data display on operator performance, the majority of these guidelines are based on convention and general aesthetics rather than formal experimentation. A useful overview of these guidelines is shown in Table 1. These represent the primary issues one should consider when involved in designing effective screens for data display.

TABLE 1. OVERVIEW OF GUIDELINES FOR DISPLAY OF DATA (Adapted from References [135] and [53])

Display

- Data displays must always be interpreted in the context of task requirements and user expectations.
- A means must be found to provide and maintain context in data displays so that the user can find the information he/she needs for the job. Task Analysis may point the way here, indicating what data are relevant to each stage of task performance.

Consistency

- Design guidelines must emphasize the value of displaying no more data than the user needs, maintaining consistent display formats so that the user always knows where to look for different kinds of information, using consistent labeling to help the user relate different kinds of information, and using consistent labeling to help the user relate different data items, on any one display and from one display to another.

TABLE 1 (*continued*)

Flexibility

- Flexibility is needed so that data displays can be tailored on-line to user needs.

Data Entry/Retrieval

- In tasks where a user must both enter and retrieve data, which is often the case, the formatting of data displays should be compatible with the methods used for data entry. Display design should also be compatible with dialogue types used for sequence control and with hardware capabilities [135].

Brevity

- Limit the screen (or transaction) to that information necessary to perform actions, make decisions, or answer questions.

Organization

- For multiple-screen transactions requiring searching through several screens, locate the most frequently requested information on the earliest transaction screens.

Layout

- Do not pack the screen with information. Use spaces and lines to perceptually organize the screen in a balanced manner.
- Group information in a logical and orderly manner with the most frequently requested information in the upper left-hand corner.
- Columnize, maintaining a top–to–bottom, left–to-right orientation;.
- Information contained on an inquiry screen should only be that which is relevant.
- On inquiry screens, the displayed data should be emphasized, since this is what the experienced user is scanning, usually by context. In looking for a date, for instance, a person's visual search usually involves scanning for numeric characters in a certain structure (such as 09/21/63), while a name search might involve scanning for a recognizable combination of alphabetic characters of an approximate size and format (such as "Johnson, Carl") [37].

Data Entry

Data entry refers to the process whereby users input data into the system. The majority of guidelines for data entry tend to emphasize input field size, layout, and design for operator transaction. Data entry screens should be designed to collect information quickly and accurately. Galitz [53] noted that the most important variable in data entry screen design is the availability of a specially designed source document from which data are keyed. This distinction is an important consideration because it determines whether keying aids are built into the screens or into the source document.

Guidelines	Research Support	Source
1. When form filling, the user should be allowed to RESTART, CANCEL, or BACKUP and change any item before taking a final ENTER action.	L	135
2. Whenever possible, multiple data items should be entered without the need for special separators or delimiters, either by keying into predefined entry fields or by including simple spaces between sequentially keyed items.	L	135, 53
3. When a field delimiter must be used for data entry, a standard character should be adopted for that purpose; a slash (/) is recommended.	L	135
4. For all dialogue types involving prompting, data entries should be prompted explicitly by displayed labels, for data fields and/or by associated user guidance messages; e.g., NAME:_ _ _ _ _ _ _ _ _ ORGANIZATION: _ _ / _ _ PHONE: _ _ _ -_ _ _ _	L	135

Data Field Guidelines

5. Field labels, should consistently indicate what data items are to be entered.	L	135
6. In ordinary use, field labels, should be protected and transparent to keyboard	L	135

control, so that the cursor skips over them when spacing or tabbing.

7. Special characters should be used to delineate each data field; a broken-line underscore is recommended;[a] e.g., L 135

   Enter plant code:_ _ _ _ _ _ _ _ _ _

8. Implicit prompting by field delineation should indicate a fixed or maximum acceptable length of the entry;[b] e.g., L 135, 18

   Enter ID: _ _ _ _ _ _ _ _ _ _

9. Input prompts should indicate which entries are mandatory and which are optional. Mandatory fields should be located ahead of optional ones. All inputs should be mandatory unless they are marked optional. L 18

   Enter plant code: _ _ _ _ _ _ _ _ _ _ _ _ _
   Region (optional): _ _ _ _ _ _ _ _ _ _ _ _ _
   Architectural Engineer (optional):_ _ _ _

10. When item length is variable, the user should not have to justify an entry either right or left and should not have to remove any unused underscores; computer processing should handle those details automatically. L 135, 53

11. When multiple items (especially those of variable length) will be entered by a skilled touch typist, each data field should end with an extra (blank) character space; software should be designed to prevent keying into a blank space, and an auditory signal should be provided to alert the user when that happens.[c] L 135

12. Labels for data fields should be distinctively worded so that they will not be readily confused with data entries, labeled control options, guidance messages, or other displayed material. L 135

13.	When displayed data forms are crowded, auxiliary coding should be adopted to distinguish labels, from data.[d]	L	135
14.	In labeling data fields, only agreed terms, codes, and/or abbreviations should be used.[e]	L	135, 40
15.	The label for each entry field should end with a special symbol, signifying that an entry may be made;[f]	L	135, 53

A colon is recommended for this purpose, e.g;

NAME: _ _ _ _ _ _ _

16.	Labels for data fields may incorporate additional cueing of data formats when that seems helpful; e.g.,	L	135

DATE (M/D/Y): _ _ / _ _ / _ _

DATE : _ _ / _ _ / _ _

M M D D Y Y

17.	When a measurement unit is consistently associated with a particular data field, it should be displayed as part of the fixed label rather than entered by the user. (data should be keyed without dimensional units, e.g., gpm, psig, Klb/hr, etc.)	L	135, 53

HIGH PRESSURE INJECTION (GPM): _ _ _

MAIN FEED CONTROL (PCT OPEN):_ _ _ _

Data Familiarity and Sequence

18.	Data should be entered in units that are familiar to the user.[g]	L	135, 18
19.	When data entry involves transcription from source documents, form-filling displays should match (or be compatible with) paper forms; in a question-and-answer dialogue, the sequence of entry should match the data sequence in source documents.[h]	L	135
20.	If no source document or external information is involved, the ordering of multiple-item	L	135

	data entries should follow the logical sequence in which the user is expected to think of them.[i]		
21.	When a form for data entry is displayed, the cursor should be positioned automatically in the first entry field.[j]	L	135
22.	When sets of data items must be entered sequentially in a repetitive series, a tabular format where data sets are keyed row by row should be used.[k]	L	135
23.	Justification of tabular data entries should be handled automatically by the computer; the user should not have to enter any leading blanks or other formatting characters; e.g., if a user enters 56 in a field four characters long, the system should not interpret 56__ as 5600.	L	135, 53, 18
24.	It should be possible for the user to make numeric entries (e.g., dollars and cents) as left-justified, but they should be automatically justified with respect to a fixed decimal point when a display of those data is subsequently regenerated for review by the user.	L	135
25.	For dense tables (those with many row entries), some extra visual cue should be provided to guide the user accurately across columns.[l]	L	135
26.	Software for automatic data validation should be incorporated to check any item whose entry and/or correct format or content is required for subsequent data processing.[m]	L	135
	Transactions, Cross Referencing, and Flexibility		
27.	In a repetitive data entry task, data validation for one transaction should be completed and the user allowed to correct errors before another transaction begins.[n]	L	135
28.	When helpful values for data entry cannot be predicted by user system interface (USI) designers, which is often the case, the user (or perhaps some authorized supervisor) should have a special transaction to define, change, or remove default values for any data entry field.	L	135

29.	On initiation of a data entry transaction, currently defined default values should be displayed automatically in their appropriate data fields.[o]	L	135
30.	User acceptance of a displayed default value for entry should be accomplished by simple means, such as by a single confirming key action or simply by tabbing past the default field.[p]	L	135, 53
31.	A user should not be required to enter bookkeeping data that the computer could determine automatically.[q] For example, a user generally should not have to identify his workstation to initiate a transaction, nor include other routine data such as transaction sequence codes.	L	135
32.	A user should not be required to enter redundant data already accessible to the computer. The user should not have to enter such data again.[r] For example, the user should not have to enter both an item name and identification code when either one defines the other.	L	135
33.	Whenever needed, automatic cross-file updating should be provided so that a user does not have to enter the same data twice.	L	135
34.	When data entry requirements may change, which is often the case, some means should be provided for the user or supervisor to make necessary changes to data entry procedures, entry formats, data validation logic, and other associated data processing.	L	135
35.	Areas of the screen not containing entry fields (i.e., protected fields) should be inaccessible to the operators and not require repeated key depressions to step through.	L	53
36.	Space lines should be incorporated where visual breaks or spaces occur on the source document.	L	53
37.	A section heading should be located directly above its associated data fields.	L	53

38. When possible, stacking of input or multiple entries should be permitted;[s] e.g., L 47, 18

Code	Category
LXX	Labor
MXX	Material

Selection Code: LXX; LRX; LRT (The last two codes normally are displayed at the next two menu levels, but three entries are stacked here to save time.)

39. The user should be able to alter input during and after entry.[t] L 18

40. In a variable-length entry, the user should be required to enter only the relevant input data.[u] L 18

41. When possible, a system should recognize common misspellings of a command and execute the command as if it had been spelled correctly.[v] L 18

42. Misspelling of similar commands should not cause errors.[w] L 18

43. Keying should be minimized.[x] L 18

44. The user should not be required to reenter parameters that have not changed since the previous interaction.[y] L 18

---

[a] Implicit prompts help reduce data entry errors by the user [135].

[b] Prompting by delineation is more effective than simply telling the user how long an entry should be. Underscoring gives a direct visual cue as to the number of characters to be entered, and the user does not have to count them. Similar implicit cues should be provided when data entry is prompted by auditory displays. Tone codes can be used to indicate the type and length of expected data entries [135], [18].

[c] Guideline11 permits consistent use of tab keying to move from one field to the next [135].

[d] For novice users, it may sometimes be helpful to have brighter labels, if that could be provided as a selectable option [135].

[e] Do not create new jargon; if in doubt, pretest all proposed wording with a sample of qualified users [135], [18].

[f] A symbol should be chosen that can be reserved exclusively for prompting user entries or else is rarely used for any other purpose [135], [53].

[g] Data conversion, if necessary, should be handled by the computer [135], [18].

[h] When paper forms are not optimal for data entry, consider revising the layout of the paper form. When data entries must follow an arbitrary sequence of external information (e.g., keying telephoned reservation data), some form of command language dialogue should be used instead of form filling to identify each item as it is entered so that the user does not have to remember and reorder items [135], [53].

[i] Alternately, data entry can sometimes be made more efficient by placing all required fields before any optional fields [135].

[j] As an exception to Guideline 21, if a data form is regenerated following an entry error, the cursor should be positioned in the first field in which an error has been detected [135].

[k] As an exception to Guideline 22, when the items in each data set will exceed the display capacity of a single row, tabular entry will usually not be desirable. Row-by-row entry will facilitate comparison of related data items and permit potential use of a DITTO key for easy duplication of repeated entries [135].

[l] A blank line after every fifth row is recommended. Alternatively, adding dots between columns at every fifth row may suffice. This practice is probably more critical for accurate data review and change than it is for initial data entry, but it is desirable in the interest of compatible display formats [135].

[m] Do not rely on the user always to make correct entries. When validity of data entries can be checked automatically, such computer aids will help improve accuracy of data entry. Some data entries, of course, may not need checking or may not lend themselves to computer checking, such as free text entries in a COMMENT field [135].

[n] Guideline 27 is particularly important when the user is transcribing data from source documents, so that detected entry errors can be corrected while the relevant document is still at hand [135].

[o] The user should not be expected to remember default values. It may be helpful to mark or highlight default values in some way to distinguish them from new data entries [135], [53].

[p] Similar techniques should be used in tasks involving user review of previously entered data [135], [53].

[q] Complicated data entry routines imposed in the interest of security may hinder the user in achieving effective task performance; other means of ensuring data security should be considered [135].

[r] As an exception to Guideline 32, redundant data entry may be used for resolving ambiguous entries, for user training, or for security (e.g., user identification). Verification of previously entered data is often better handled by review and confirmation rather than by reentry [135].

[s] When an experienced user is responding to the first of a sequence of selection screens and he knows before seeing them how he will respond to the next several screens, he should be able to input several entries at the same time. Sequential entries can be separated by a special character; a semicolon is recommended [47], [18].

[t] For instance, a system should allow erasure or cursor repositioning to overwrite previously entered input. The input specifications should remain on the screen when the requested data are displayed so the user can easily alter portions of the input without having to re-enter all of it to generate a slightly different subset of the requested data [18].

[u] Do not require the user to erase field length indicators (e.g., erase underscores), to right- or left-justify the entry, etc. If the entry is shorter than the maximal field length, its position in the field should not matter [18].

[v] Permitting abbreviated commands can also serve this function by ignoring all noncritical characters in a command. Another method is to equate selected anticipated misspellings to the correctly spelled command [18].

[w] While the system should be tolerant of common misspellings, spelling errors should not produce valid system commands or initiate processing sequences which are different from those intended. Consider possible confusions with existing commands when selecting new command words [18].

[x] Make input keywords short and have them approximate real words to minimize the amount of typing required. When possible, allow the user to select the item number rather than require the entry of a longer item code or words [18].

[y] Also avoid requirements for user entry of information already available to the system, such as the current date.

Comment. A useful overview of these guidelines is shown in Table 2.

TABLE 2. OVERVIEW OF GUIDELINES FOR ENTRY OF DATA (Adapted from References [53] and [18].)

Input of data

- Entries should be made by selecting from a list of digits (e.g., from a call-up help screen) rather than by a typed command to minimize opportunities for typing errors [18].

Field Designation

- Where to key data should always be obvious, ideally through the use of underscores used to define field sizes. If underscores are not possible, the use of a caption symbol (such as the colon) to signify the starting point of an entry field is acceptable.
- The entry field preceded by one blank space will usually start immediately after a symbol.
- Absolute identification of field length is not essential if the system is operating with manual tabbing; the terminal will identify the end by locking the keyboard.

Source Document

- With horizontally arranged fields using a source document, it is preferable to leave three blank spaces between the last character of one field and the caption of the next. If space constraints exist, one blank space is acceptable.
- Vertical spacing between rows of fields on a screen will follow the spacing conventions of the source document. This is another method of maintaining an image relationship between document and screen [53].

The extensive set of guidelines tends to emphasize the design of effective input procedures. This is accomplished in the guidelines by making the operator/computer interface accessible to both skilled and novice users. Adherence to these guidelines should provide an interface that will minimize the potential for human error and time to enter data.

Instructions – General

This section examines the techniques, methods, and guidelines for presentation of instructional materials on a CRT screen. These guidelines are also applicable to presenting CRT-based procedures and for determining information requirements. This section is highly related to variables previously examined, such as messages and text, and the reader should review those sections for further detail.

Guidelines	Research Support	Source
1. Words in instructions should be meaningful to the user.[a]	Y	5, 42
2. Short words should be used in instructions.[b]	L	5
3. Active voice and the affirmative case should be used in instructions.[c]	Y	5, 42
4. Instructions should be patterned in a logical consistent manner.[d]	Y	5

[a] Word meaning is related to such variables as word frequency, overall familiarity, complexity and user's lexicon. The meaningfulness of words is usually defined in terms of their associative value. Meaningful words generally evoke greater imagery and understanding than less meaningful words [5].

[b] Preference should be given to the use of words containing few syllables. However, it is better to use multisyllable words when they can convey the intended meaning more readily than words with fewer syllables [5].

[c] As a general rule, positive (or affirmative) and active sentences are the easiest to understand. Introducing the passive or the negative creates problems, either by slowing people down or by causing them to make errors [5].

[d] Patterning refers to the use of underlining, boldface type, uppercase, or any other techniques to emphasize certain words in order to stress their importance. Patterning provided by capitalizing the word or words that are most important in each sentence tends to help people comprehend more [5].

Instructions–Illustrations

Illustrations generally should not be included if they repeat in some graphical form what has been said adequately in words. Thus, the use of illustrations should depend on the illustration adding something to the meaning of an instruction. More specifically, illustrations should be used to show details that are difficult to describe verbally. For example, figures generally present detailed numerical information in a compact and organized way. Probably the best approach to incorporating illustrations in instructions is to determine the illustrations first and then write around the illustrations with whatever narrative is then needed [5].

	Guidelines	Research Support	Source
1.	Illustrations should be appropriate for the type of information to be conveyed.[a]	L	5
2.	Illustrations should be placed close to the corresponding text.[b]	L	5
3.	Wording on illustrations should be minimized.[c]	L	5
4.	Tables and graphs should be captioned.[d]	L	5

[a] One of the most difficult problems with providing adequate illustrations in a set of instructions is determining the most appropriate illustration for a particular type of information. Tables work well when large amounts of specific numerical data must be presented. (See Data Display for additional guidelines.) [5].

[b] Illustrations should not be saved up and then presented as a set at the end of a text. The only exception might be when an illustration needs to be consulted throughout a set of instructions [5].

[c] A designer should seek to keep wording on an illustration to a bare minimum. In addition, illustrations should not be overloaded with data or have multiple subjects covered on a single display [5].

[d] Tables and graphs communicate most effectively when they are accompanied by good titles or captions. Titles should be placed above tables and below graphs or other figures. Tables and graphs particularly should be numbered and have titles that are descriptive of the content. In addition, words used in titles and labels of illustrations should be consistent with the

words used in the text. Titles should be short; five and eleven words tend to be remembered best. If a subtitle is used to help clarify the title, the subtitle should be placed beneath the title and have lettering that is smaller than the size of the title letters [5].

Instructions–Style and Structure

The manner in which instructions are presented can significantly influence readability and comprehension of the material. Various parameters such as introductions, headings, titles, topic sentences, and summaries can all be used to enhance the effectiveness of instructions [5].

	Guidelines	**Research Support**	**Source**
1.	When instructions must be rapidly accessed, a table of contents and/or an index should be provided.[a]	L	5
2.	The literary style of a set of instructions should be appropriate to its intended use.[b]	L	5
3.	Instructions should have a clearly stated beginning and a well-developed summary.[c]	Y	5
4.	Paragraphs of text should be short and should contain a single idea.[d]	L	5
5.	Instructions should be simple.[e]	L	5
6.	Instructions should state important items more than once.[f]	L	5
7.	Instructions should contain only essential information.[g]	L	5
8.	The amount of detail should be appropriate to the experience of the user.[h]	L	5
9.	The sequence of the instructions should follow the sequence of actions required.	L	21
10.	Short sentences, flow diagrams, algorithms, lists, and tables are superior to prose.	Y	5, 42

11.	The main topic of the instruction should appear at the beginning of the sentence.	Y	5, 42
12.	All instructions should be tested on naive users before being finalized.	Y	42
13.	Many-step instructions should use a two–column format.[i]	L	42
14.	In a list of specifications for service or supply, more than a part number should be given.[j]	Y	42
15.	Warning and caution notices should be accurate and concise and should contain only the information relevant to the warning or caution. They should not contain operator actions.	L	152
16.	Warnings and cautions should immediately precede the steps to which they refer.	L	152

---

[a] Where it is necessary for a set of instructions to be rapidly accessed, a table of contents or index or both should be included. It may be useful for both the traditional hierarchical and the keyword types of indexing to be included [5].

[b] Probably the most frequently used approach is a narrative one in which instructions are presented in a chronological sequence. The narrative approach is obviously very useful for procedural instructions, although it frequently does not allow a designer to show the relative importance of various items [5].

[c] Material at the beginning and end of a set of instructions tends to be learned most rapidly and recalled more easily. This places great emphasis on a clearly stated beginning for a set of instructions and, possibly, a well-developed summary at the end [5].

[d] Shorter paragraphs tend to facilitate comprehension better than longer paragraphs [5].

[e] Simplify the presentation of instructions to their essential point, to avoid unnecessary detail and to maximize the ability to identify and understand critical material [5].

[f] Dialog may be achieved in two ways–by repetition of the same material or by providing alternative ways of looking at the same information (e.g., with examples) [5].

g To communicate instructions effectively, the designer must be very selective of the information available. Attempting to transmit all available information to a user would in most systems produce only confusion [5].

h Experienced users perform better when given more general instructions, while users with less experience seem to derive more benefit from very specific instructions. A general type of instruction may be preferable in situations demanding originality or creativity on the part of a user [5].

i The two-column format is an appropriate design for many-step instructions. The second column (Further Information) is especially useful during an operator's learning phase on the equipment.

j For instance, the specification should be supplemented with a short description (MU32 – Make Up Pump), which relieves the operator from memorizing the alphanumeric code when following the instructions. If there are any hazards to warn the operator about, highlight the warnings and describe the potential consequences of not following the instructions.

Comment. Bailey [5] provides a fairly comprehensive review of the guidelines for effective presentation of instructions. A sizable proportion demonstrate evidence of formal research support, while others rely on general human engineering practices and convention. The research support has also been gleaned from hardcopy experimentation and not on CRT-generated displays. However, it is not clear how this factor would compromise the guidelines application to CRT-presented information. The guidelines are also highly subjective as directly usable data for assessing instructional information. A certain level of interpretation is required. These limitations, as pointed out previously, are not imposed by lack of available research. Rather, these limitations are due to the difficulty of developing a comprehensive set of generic guidelines suitable for all tasks. That is, instructions must be custom-developed in most cases. From the design perspective, the following questions should be answered before the guidelines can be applied:

a Who are the users?

b What is the relative importance of each instruction?

c What is the proper location for the instructions?

d How much information redundancy is appropriate?

e What is the most effective method and format of presentation (diagrams, photographs, prose)?

f How intelligible are the instructions?

g Is the appropriate nomenclature used?

Characteristics of Alphanumeric Characters

This subcategory examines the various dimensions pertaining to characters and character sets that contain letters and/or numbers. Guidelines for determining maximally legible alphanumerics are classified and discussed under the following variables:

- Font or style
- Character size and proportion
- Character case

Font or Style

Font is the typeface used, which encompasses several styles of type and refers to the fundamental geometry of the letters and numbers. Fonts, or typefaces, are generally referenced by standard names and by the type size (in points)–for example, Roman 12 pt or 8 pt agate. The number of actual fonts available on CRTs is unlimited.

	Guidelines	Research Support	Source
1.	In designing a visual display character set, each character should be designed so that fine differences in stroke length, curvature, etc., are preserved in order to avoid similarity.	Y	23
2.	For a given font, it should be possible to clearly distinguish between the following characters:	Y	23

X and K	I and L	O and Q	U and V
T and Y	I and 1	S and 5	

Comments. The basic evaluation criterion for font selection should probably be legibility and, for most computers, is more influenced by resolution, character size, spacing, and interline spacing than by font styles. Preference should be given to simple styles with straight lines and clear differences between O and 0 (zero) and between S and 5. Script and other highly stylized fonts should be avoided (e.g., shadow, calligraphy).

Character Size and Proportion

Both character size and proportion are indexed to the height of the characters. Standard sizes are measured in millimeters, fractions of an inch, or else in the visual angle subtended at some given viewing distance. The direct linear measurements are meaningful only in terms of a fixed viewing distance, usually the standard reading distance of 18 to 24 in. To maintain a stable viewing angle, the size of the characters must change as the viewing distance changes. The linear measurement of the major dimensions of characters can be accomplished with a ruler (preferably a millimeter rule). These measures can be converted to visual angle by use of the following formula: $\alpha = \tan^{-1} H/D$, where α is the visual angle, H is the character dimension (in cm.), and D is the viewing distance (in cm.). This formula introduces a slight error; the correct formula would take into account that the character dimension is bisected by the line of sight, but the error is less than rounding error for most cases. This relationship is illustrated in Figure 6.

Because the apparent visual size of objects changes as a function of viewing distance, the standards must reference some stable unit. The unit of choice in visual size has been the height of the object, with all other dimensions referenced to it. Standards for height are based on capital letters.

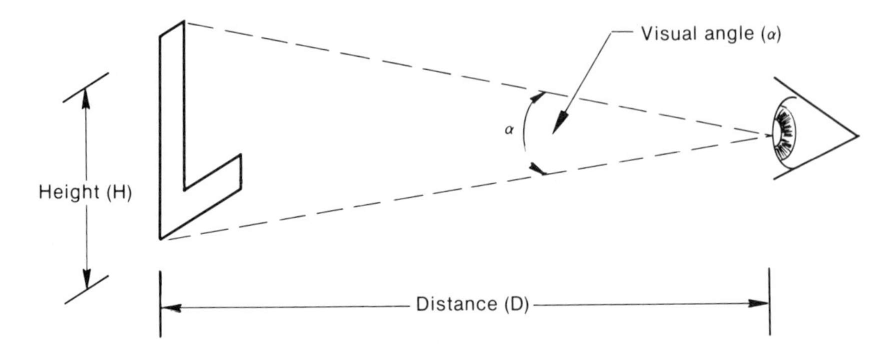

Figure 6. Visual angle as a function of distance and character height.

Guidelines		**Research Support**	**Source**
1.	Character height should be 16 minutes of arc to 26.8 minutes of arc, with 20 minutes of arc preferred.[a]	Y	146, 23, 22, 41, 58, 132, 45
2.	The ratio between character height and width should be from 1:1 to 5:3.[b]	Y	150, 90, 147, 146
3.	The ratio between character height and stroke width should be from 5:1 to 8:1.[c]	Y	150, 129, 11, 48

4.	The minimum spacing between characters should be one stroke width.[d]	Y	150, 23, 22, 41, 147
5.	The minimum spacing between words should be one character width.	Y	150, 147
6.	Spacing between lines should be from 50 to 150% of the character height.[e]	Y	150, 23, 22, 49

[a] Reference [41] recommends 18 min of arc for viewing distances greater than 50 cm and 2.6 mm minimum for viewing distances less than 50 cm. Reference [58] recommends a minimum of 3.5 mm. In Reference [23], 2.3 to 29 mm, depending on viewing distance, is recommended. Reference [146] recommends 16.4 min of arc to 26.8 min of arc, depending on contrast level for a viewing distance of 16 inches. References [132] and [45] suggest 20 min of arc for luminance levels of 0.01 to 10 ftL and 10 min of arc for luminance levels of 10 to 50 ftL.

[b] References [150] and [147] note as exceptions the characters M, W, and 4, which should have a height-to-width ratio of 5:4, and 1, which should be one stroke wide. Reference [146] recommends a height-to-width ratio of 4:3.

[c] References [150], [11], and [12] recommend a height-to-stroke width ratio from 6:1 to 8:1. Reference [129] recommends a ratio of 5:1.

[d] References [150] and [147] recommend a minimum of one stroke width between characters. In Reference [22] a minimum of one-half character height between characters is recommended. Reference [41] specifies a minimum of 10% of the character height between characters. Reference [22] specifies from 20 %to 50% of character height between characters.

[e] Reference [150] recommends from 10% to 65% of character height between lines (considering ascenders and descenders). Reference [22] recommends one-half of character height between lines. Reference [121] recommends one character height between lines. Reference [23] recommends from 100% to 150% of character height between lines.

Comment. Character size and spacing must be evaluated in the context of the purpose of the display. That is, labels for displays or isolated words in menus can be best seen and understood when set in larger size with greater spacing, while long paragraphs of text are most easily read and understood with smaller size and closer spacing [139], [83]. The legibility of closely spaced characters is also affected by use of upper and lowercase (see below).

Character Case

Character case refers to the use of different styles of upper and lower (capital and small) alphabetical symbols. This subject has been very poorly researched for CRT displays; the one relevant standard [144] assumes that the general reading research is relevant.

Guidelines	Research Support	Source
1. Labels or statements should be in uppercase.	Y	18
2. Text should be displayed in both uppercase and lowercase.	Y	18

Comment. Pending future research on the effects of character case on human performance, labels, and short messages should be all uppercase, while text should be mixed case.

Screen Organization and Layout

The screen designer is limited in the ability to specify the size of the screen needed to effectively present information to the user. Therefore, given a set screen size (which is almost never large enough), the designer is tasked with the function of fitting information on the screen in a manner that ensures easy assimilation by the user. That is, the mode for organizing and layout of information must be in a form appropriate to satisfy the task requirements of the user. The following sections address the various techniques employed to organize information on the screen. The specific structures for the optimal presentation of information to the user have been described previously.

In this subcategory, the following variables are examined:

- Screen size
- Grouping

- Display density
- Display partitioning/windows
- Frame specifications
- Interframe considerations – paging and scrolling
- Interframe considerations – windowing

Screen Size

The physical size of the CRT screen is examined in this section. The value for area is usually specified in terms of the diagonal size i.e., corner-to-corner dimension of the screen or by the height and width dimensions of the viewing area [23]. Diagonal size and height and width dimensions are shown in Figure 7. Most desk-top screen sizes have a diagonal dimension of 12 or 15 inches.

Another common measurement is the aspect ratio of the screen. This is expressed as the ratio between the height and width of the viewing area. An aspect ratio of 3:4 is most commonly encountered in horizontally mounted CRT screens. The units of measure can be inches, centimeters, and in some cases the number of pixels. A distinction should be made between screen size and viewing area. Viewing area is always somewhat less than the size of the screen; manufacturers will vary this difference for one or more of the following reasons: (a) to provide as large a viewing area as possible; (b) to minimize distortion at the edges of the screen; and (c) to achieve a particular aspect ratio. Screen designers should be aware of this minor distinction when specifying a particular screen size.

Screen size affects the interaction with the operator and the perceived resolution of the display. In an absolute sense, resolution also affects the required bandwidth or data throughput rate of the monitor and interface circuits and, ultimately, the size and cost of the display memory, although these will already have been taken into account in the design of a complete colorgraphics display system [110].

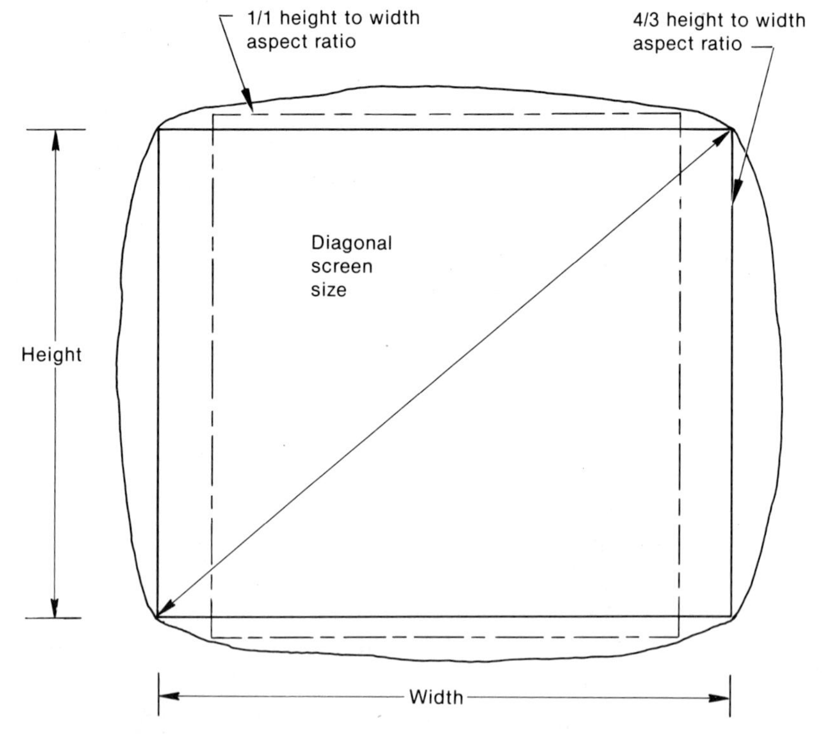

Figure 7. Display dimensions for various CRT screen sizes (Source: Reference [110]).

The principal advantage to be gained by increasing the size of the screen is simply a larger spatial distribution of the display [110]. However, too large a display may cause problems with flicker, sharpness of image, colors registration, glare, etc. [18].

Guidelines	Research Support	Source
1. The screen should be the smallest size which will allow required information to be seen clearly and easily by the viewer.[a]	N	18, 110
2. The screen should take into account the distance of the operator from the screen (e.g., large screen overviews).	L	Authors

[a] When too small, smaller screen sizes compact image elements and may cause overlap and loss of information. The consensus of human factor principles call for the smallest size which will allow required information to be seen clearly and easily by the viewer [110].

Comment. A survey of the literature reveals no specific guideline set for screen size. This is probably due to the task dependency requirements for determining screen size. The limited guidelines in this area tend to regard screen size as more a function of viewing distance, visual angle, and visual arc of the characters/symbols displayed on the screen. MIL-STD-1472C [146], Section 5.2.5, discusses guidelines pertaining to "Large Screen displays," but no specific dimensions are discussed, i.e., how big is an acceptable large screen size? To summarize, the guidelines indicate that too large a screen size is not practical, but the screen should not hinder legibility by being too small. Recent trends indicate, at least for nuclear control rooms in Japan, utilization of a large screen overview that is accessed by an entire crew. Little data exist which demonstrate a clear performance advantage with this approach.

Grouping

Within the context of screen organization, grouping refers to the placement of data into functional or otherwise meaningful sets [9]. There are generally four techniques for organizing data: (a) sequence, (b) frequency, (c) function, and (d) importance. The division of data into groups of data or information fields is necessary because the human information processing mechanism has limits on the number of items it can efficiently process at one time. In a manner similar to the presentation of nontextual information, the process of grouping information on a computer display uses our knowledge of the chunking process to build displays that aid the absorption of information by the user.

	Guidelines	**Research Support**	**Source**
1.	Information that is continually being transmitted or received should be sequentially grouped.	Y	5, 135, 18
2.	Information should be grouped in the order of its frequency of use.	Y	5, 135, 18
3.	If frequency of use is not a major concern, information should be functionally grouped.	Y	5, 135, 18
4.	When some items are more critical than others to the success of the systems, the information should be grouped by importance.	Y	5, 135, 18
5.	Grouped data should be arranged in the display with consistent placement of items so that user detection of similarities, differences, trends, and relationships is facilitated.	Y	135, 18
6.	When there is no appropriate logic for grouping data (sequence, function, frequency, or importance), some other principle should be adopted, such as alphabetical or chronological grouping.	L	135, 18

7.	Similar information should be displayed in groups according to the left-to-right or top-to-bottom rules.	L	146
8.	All displayed data necessary to support an operator activity or sequence of activities should be grouped together.	L	146
9.	Screens should provide cohesive groupings of screen elements so that people perceive large screens as consisting of smaller identifiable pieces. People are relatively efficient at viewing groups or chunks of data.	L	53
10.	Providing perceptual structure on the screen can be achieved in a variety of ways, ranging from some arbitrary but consistent grouping to an optimally designed functional grouping based upon frequency of usage data.	L	47
11.	Grouping similar items together in a display format improves their readability and can highlight relationships between different groups of data. Grouping can be used to provide structure in the display and aid in the recognition and identification of specific items of information.	L	23

Comment. Bailey [5] states, "Grouping items on a video display is a delicate process of elimination, weighting, and judgment." [5]

Many of the methods for grouping are discussed in an assortment of human engineering design documents. The approaches were originally derived from guidelines for the arrangement of displays and controls on a control panel. However, there is no rationale for limiting these methods to their original application. The generalization of these techniques to CRT screens appears to be a valid assumption. Bailey's guidelines [5] are supported by the authors of several other design documents [135], [47], [18]. Bailey's guidelines are based on the concept that the designer knows what the user is going to do with the information presented on the screen. Therefore, sequential task analysis, information requirements analysis, and procedural flow analysis are necessary requirements. The data from the task and/or information analysis are also useful in determining function and importance groupings. MIL-STD-1472C [146] and Galitz [53] recommend some type of grouping strategy but fail to be any more specific.

Display Density

In most cases, screen designers are not provided the luxury of unlimited screen size. All too often, the designer tends to pack too much information on the screen at one time. At some point, the packing density of information begins to adversely affect operator performance. It is often difficult to determine just what is too much. Galitz [53] notes, "How much display information is 'too much' has not been determined. An ultimate answer must undoubtedly reflect, among other things, the requirements of the application and how the screen is formatted."

Density, in the context of screen design, is often referred to as the amount of space used expressed in a percentage of the total screen area. Displays of 30%, 50%, and 70% packing density are shown in Figure 8. There is an optimal percentage of information that can be effectively presented to the operator on a single screen. However, that optimal value is highly dependent on the specific operator's task. Danchak [36] states, "It is intuitively obvious that an upper limit exists on the amount of active screen area. Quantifying this is another matter." In the literature, the term density is often synonymous with display loading or clutter.

Guidelines	**Research Support**	**Source**
1. Screen packing density should not exceed 50% and preferably should be less than 25%.[a]	Y	150, 42, 36
2. Display screens should be perceived as "uncluttered."[b]	Y	47, 18

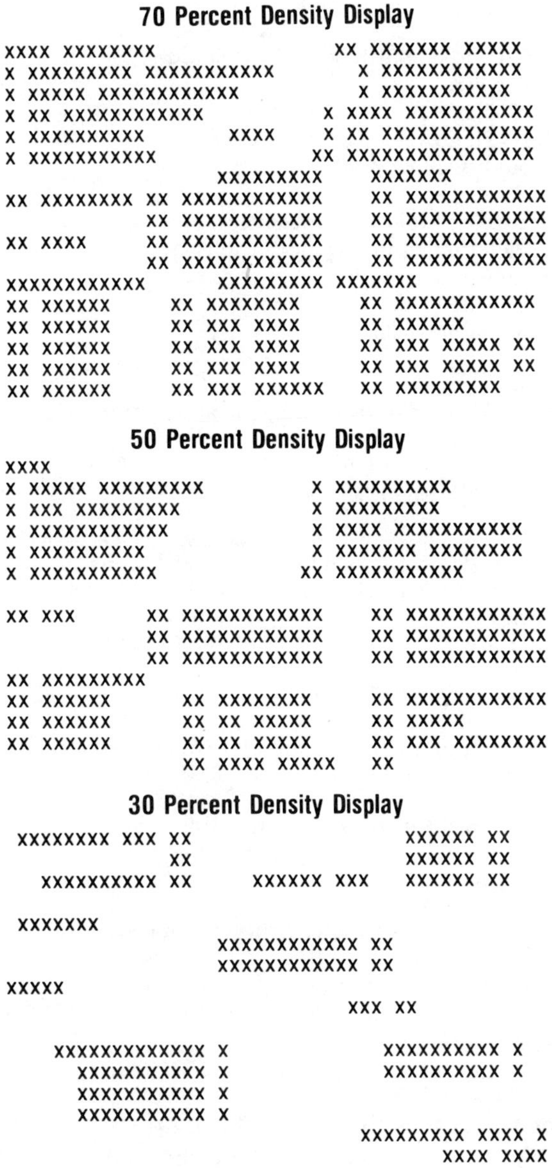

70 Percent Density Display

```
XXXX XXXXXXXX               XX XXXXXXX XXXXX
X XXXXXXXXX XXXXXXXXXXX       X XXXXXXXXXXXX
X XXXXX XXXXXXXXXXXX          X XXXXXXXXXXX
X XX XXXXXXXXXXXX          X XXXX XXXXXXXXXXX
X XXXXXXXXXX       XXXX    X XX XXXXXXXXXXXXX
X XXXXXXXXXXX             XX XXXXXXXXXXXXXXXX
                  XXXXXXXXX    XXXXXXX
XX XXXXXXXX XX XXXXXXXXXXXX    XX XXXXXXXXXXXX
            XX XXXXXXXXXXXX    XX XXXXXXXXXXXX
XX XXXX     XX XXXXXXXXXXXX    XX XXXXXXXXXXXX
            XX XXXXXXXXXXXX    XX XXXXXXXXXXXX
XXXXXXXXXXXX      XXXXXXXXX XXXXXXX
XX XXXXXX     XX XXXXXXXX     XX XXXXXXXXXXXX
XX XXXXXX     XX XXX XXXX     XX XXXXXX
XX XXXXXX     XX XXX XXXX     XX XXX XXXXX XX
XX XXXXXX     XX XXX XXXX     XX XXX XXXXX XX
XX XXXXXX     XX XXX XXXXXX   XX XXXXXXXXX
```

50 Percent Density Display

```
XXXX
X XXXXX XXXXXXXXX         X XXXXXXXXXX
X XXX XXXXXXXXX           X XXXXXXXXX
X XXXXXXXXXXXX            X XXXX XXXXXXXXXXX
X XXXXXXXXXX              X XXXXXXX XXXXXXXX
X XXXXXXXXXXX            XX XXXXXXXXXXX

XX XXX      XX XXXXXXXXXXXX    XX XXXXXXXXXXXX
            XX XXXXXXXXXXXX    XX XXXXXXXXXXXX
            XX XXXXXXXXXXXX    XX XXXXXXXXXXXX
XX XXXXXXXXX
XX XXXXXX      XX XXXXXXXX     XX XXXXXXXXXXXX
XX XXXXXX      XX XX XXXXX     XX XXXXX
XX XXXXXX      XX XX XXXXX     XX XXX XXXXXXXX
               XX XXXX XXXXX   XX
```

30 Percent Density Display

```
 XXXXXXXX XXX XX                   XXXXXX XX
              XX                   XXXXXX XX
   XXXXXXXXXX XX      XXXXXX XXX   XXXXXX XX

 XXXXXXX
                  XXXXXXXXXXXX XX
                  XXXXXXXXXXXX XX
XXXXX
                             XXX XX

    XXXXXXXXXXXXX X                 XXXXXXXXXX X
      XXXXXXXXXXX X                 XXXXXXXXXX X
      XXXXXXXXXXX X
      XXXXXXXXXXX X
                              XXXXXXXXX XXXX X
                                     XXXX XXXX
```

Figure 8. Display packing densities of 30, 50, and 70 percent (from Reference [42]).

3.	Provide information that is only essential to making a decision or performing an action.[c]	L	135, 53
4.	All data related to one task should be placed on a single screen.[d]	L	135, 53
5.	For critical task sequences, screen packing density should be minimized.[e]	Y	146
6.	Where user information requirements cannot be accurately determined in advance of interface design or are variable, on-line user options should be provided for data selection, display coverage, and suppression.	L	135

---

[a] Reference [42] notes that readability of status information or other text drops off rapidly at screen packing densities greater than 50%. Reference [36] recommends a maximum of 25%. Reference [150] excludes demarcation lines (lines separating groups of data) from determination of packing density.
Experience shows that display loading (the percentage of active screen area) should not exceed 25%. This may seem extremely low until one considers that a well-designed page of printed material has a loading of only 40%. An analysis of existing CRT displays that were qualitatively judged good revealed a loading on the order of 15%. The remaining area constitutes white space that is essential for clarity in any display. The amount of variable data on these displays never exceeded 75% of the total active area.

[b] Distribute the unused area to separate logical groups, rather than having all unused area on one side [18].

[c] Screens should provide only relevant information, because the more information, the greater the competition among screen components for a person's attention [53].

[d] One should not have to remember data from one screen to the next [53].

[e] A minimum of one character space should be left blank vertically above and below critical information with a minimum of two character spaces left blank horizontally before and after [146].

Comment. Although these guidelines are relevant for textual information, they are of little use when alternative coding techniques are implemented, (e.g., graphic displays). However, Eastman Kodak

[42] recommends the use of coded information in order to pack more data on the screen without losing the desired spacing. Other methods for the reduction of information density include (a) developing a hierarchy of information needs (routinely display information only for the higher priorities and relegate the other information to routines that can be called up as needed) and (b) using graphics to display events that are changing in time. Note the diversity in these guidelines. For example, Galitz [53] suggests methods for qualifying what information should be displayed to avoid clutter. Other sources, such as NUREG-0700 [150], place a quantitative value on screen loading. Danchak [36] even indicates a value for both static and dynamic information. As evidenced from the variety of percentage values in the guidelines, a single value is difficult to derive. The authors conclude that a single value is too restrictive. Instead, the designer should be aware of the experimental evidence which indicates a significant decrement in operator performance for screen loadings greater than 50%. Emphasis should be placed on minimizing this value (preferably less than 25%). Although these authors have no hard data to substantiate this claim, it appears that more experienced users can tolerate information-laden displays of a higher density than can novice users.

Display Partitioning/Windows

The purpose of partitioning is to organize screen information into functional groups. This section focuses on the various techniques that can be applied to enhance the logical organization of the screen elements. This includes various methods and techniques which may be applied to help the user visually perceive that the screen is structured in some logical format.

Guidelines	Research Support	Source
1. Screens should be divided into windows that are clearly perceptible to the user.[a]	L	150, 135, 47, 53

2.	On large, uncluttered screens, windows should be separated using three to five rows or columns of blank space.	L	150, 47, 53
3.	Specific areas of the screen should be reserved for information such as commands, status messages, and input fields; those areas should be consistent on all screens.	L	53
4.	When a display window must be used for data scanning, the window size should be greater than one line.	L	135
5.	The screen should not be divided into a large number of small windows.	L	53
6.	When the body of the display is used for data output, the screen should be coherently formatted and not partitioned into several small windows.[b]	L	135
7.	The number of overlapping windows should be minimized.	L	Authors
8.	If possible, program windows so that the size is expandable by the user.	L	Authors

---

[a] Reference [47] suggests several ways of increasing the user's perception of structure. On a large uncluttered screen, areas (that is, windows) may be separated by blank spaces in sufficient quantity (three to five rows and/or columns) to state unequivocally that each area is confined within those blank spaces. On smaller and/or more cluttered screens, where an excess of unused spaces cannot be tolerated, the user's perception of structures among the different areas and/or objects on the screen can be assisted by the use of a variety of techniques, including different surrounding line types (solid, dashed, dotted, etc.), line widths, intensity levels, geometric shapes, colors, numbers, and letters. Barmack and Sinaiko [10] compared the efficiency of some of these codes and recommended a maximum number of levels of coding for each code type. Barmack and Sinaiko suggest other methods of coding that may also be considered:

- Motion (2 to 10 coding levels)
- Focus or distortion (2 levels)
- Line type, dashed, solid (3 to 4 levels)
- Line length (2 to 4 levels)
- Line width, boldness (2 to 3 levels)

- Orientation, location on the display surface (4 to 8 levels).

[b] To reduce search time, Reference [46] recommends that tabular displays be broken into meaningful blocks whenever possible; however, tables should not be divided into blocks to provide the impression of organization when none exists.

Comment. The above guidelines tend to exhibit three underlying recommendations for partitioning of displays:

1. On large uncluttered screens, break the display into windows, using blank spaces of sufficient quantity.

2. If screens are cluttered with a great deal of information, which probably happens frequently, apply one of the techniques recommended by Barmack and Sinaiko [10].

3. Do not break the screen into too many small windows.

An example of a display using the guidelines for display partitioning is shown in Figure 9. As with many other guidelines, there is limited research to support these guidelines, and they are somewhat vague. (How does one determine how many windows are too many?) Many guidelines have been established more from convention and human engineering criteria for layout of instrument panels, e.g., the use of demarcation lines or "dog ears" to logically group related displays on a single panel [146].

This does not imply that these guidelines are not practical, but due caution should be exercised in their application to CRT displays. In reference to the methods suggested by Barmack and Sinaiko [10] (see footnote a), methods 3 through 6 seem to be the most appropriate. Method 1 (motion), with an upper boundary of 10 levels of motion, may be unreasonably high for graphic displays of process systems and might contribute to perceptual clutter, vertigo, or other problems.

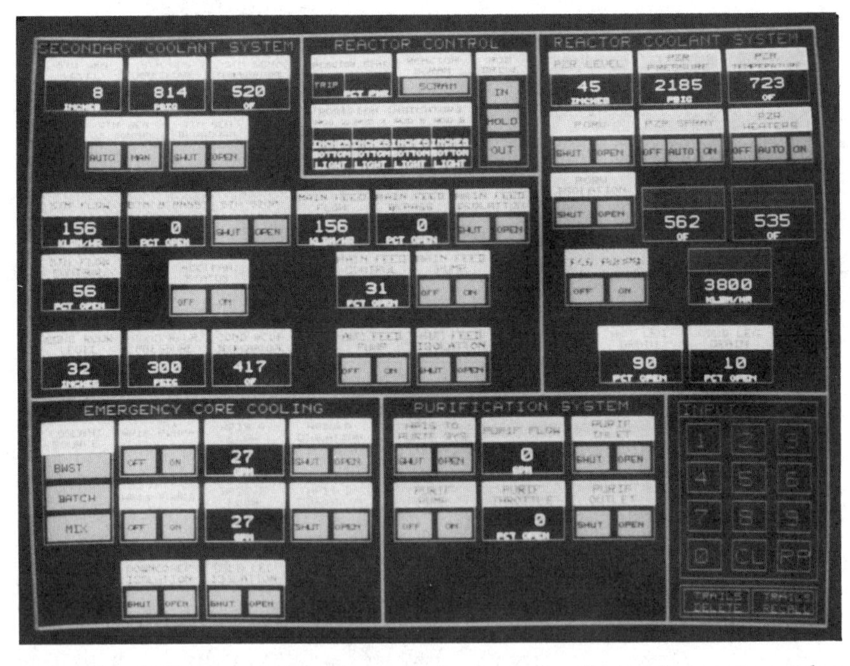

Figure 9. CRT-generated control panel operated with touch screen overlay. Windowing and demarcation lines have been implemented to partition the display into an efficient and usable configuration.

Frame Specifications

After information requirements have been defined for a particular screen or set of screens and a grouping strategy for placement of the information has been defined, the next question is usually "Where should I place the information on the screen?" The location requirements will, in most cases, be task-specific and a function of human/computer dialogue. That is, information will be placed on the screen as the dialogue scenario develops. Nevertheless, there are some general guidelines in the literature which identify where information should be placed to best facilitate the operator. This section also includes specific guidance for the placement of invariant information (common to all screens, e.g., title, page number, etc.) and, to some degree, what that invariant information should be and how it should be structured.

Guidelines	Research Support	Source
1. Specific areas of the screen should be reserved for information such as commands, status messages, and input fields; those areas should be consistent on all screens.	Y	47, 18, 54
2. Both the items on display and the displays themselves should be standardized.[a]	Y	150, 5, 146
3. An invariant field, including the page title, an alphanumeric designator, the time, and the date, should be placed at the top of each display page.[b]	Y	18, 46
4. The last four lines (at least) of each display page should be reserved for variant fields.[c]	Y	67, 18, 46
5. Procedures for user actions should be standardized.[d]	Y	18
6. Each display frame should have a unique identification (ID).[e]	Y	18

7.	Every frame should have a title on a line by itself.	Y	67
8.	Status information should be displayed near the top-right corner of the screen.[f]	Y	42
9.	Location coding should be employed to reduce operator information search time.[g]	Y	146

---

[a] Two kinds of standardization are important: standardization of items on a display and standardization of displays. Although standardization is desirable, it should not take precedence over the grouping principles of frequency, sequence, locations, and importance [5].

[b] Invariant fields, or fields that contain the same type of information for each display page, should include the page title, an alphanumeric designator, and the time and date. Such fields indicate header information and should be presented at the top of each page. Ensure that the page title indicates the purpose of the display, is in a consistent location on each page, and is separated by at least one blank line from other information. The alphanumeric designator provides a convenient means for accessing display pages, identifying them, referring to display pages in discussions, and reporting problems in page formats. Ensure that the designator is meaningful enough to be easily learned and remembered, as well as being compatible with the designation scheme applied to hardwired instruments on panels [46].

[c] Variant fields are fields reserved for information that may not be present on a given page. (Reference [40] refers to variant fields as "functional fields.") When such information is displayed on a page, the assigned screen location should always be used. Variant information might include alarm messages, data source identification, the page number (e.g., Pn of N) for consecutive pages, system messages (e.g., work orders, standby subsystems), error messages, response entry prompts, and program messages. Help identify required information for variant fields by guidelining information requirements and consulting people who are representative of operational personnel. Then prepare recommended specifications for the location and format of variant field information.[46].

[d] LOG ON and LOG OFF procedures, menu selection techniques, user input procedures, and error correction procedures are examples of user actions for which standardized conventions are required [18].

[e] The unique screen ID provides a convenient means for

- Identifying which screen is being displayed,
- Requesting display of a specific screen,

- Referring to a screen in discussions,
- Reporting trouble or problems in formats. [18]

[f] Display status information at the top of the screen, toward the right side. The location of this information will vary according to the type of operations performed, but it should be found in the same general part of the screen across all terminals in a manufacturing system [42].

[g] It has been shown that search times are significantly faster than the average for targets in the upper-right quadrant and slower for the lower-right quadrant. No difference exists for the left two quadrants. These findings can be used to advantage by placing the most important information in the upper-right quadrant and least important in the lower-right quadrant [36].

Comment. The guidelines which address information location are variable in level of detail. Certain guidelines emphasis some level of standardization between similar screens. Other guidelines extend this standardization to the point of specifying which line should be dedicated to each element. It is the opinion of the authors that many of these latter guidelines are somewhat overly restrictive. They were also laid out by a designer with a specific system in mind. Therefore, there is little potential for adapting them to a generic situation. Perhaps the most consistent observations from these guidelines are that (a) some mode of consistency and standardization should be adopted for the elements across all screens, i.e., location coding, and (b) certain pieces of guidance information should be placed on all screens. An example of a standardized format is shown in Figure 10. More research is needed before prioritizing information according to screen quadrants. The reader is further cautioned that there is a trade-off to be determined when designating invariant fields, such as titles, page numbers, and the like. When screen size is restrictive and resolution low, there is a trade-off between the quantity of data which can be presented and the space restriction that results from overreliance on invariant fields.

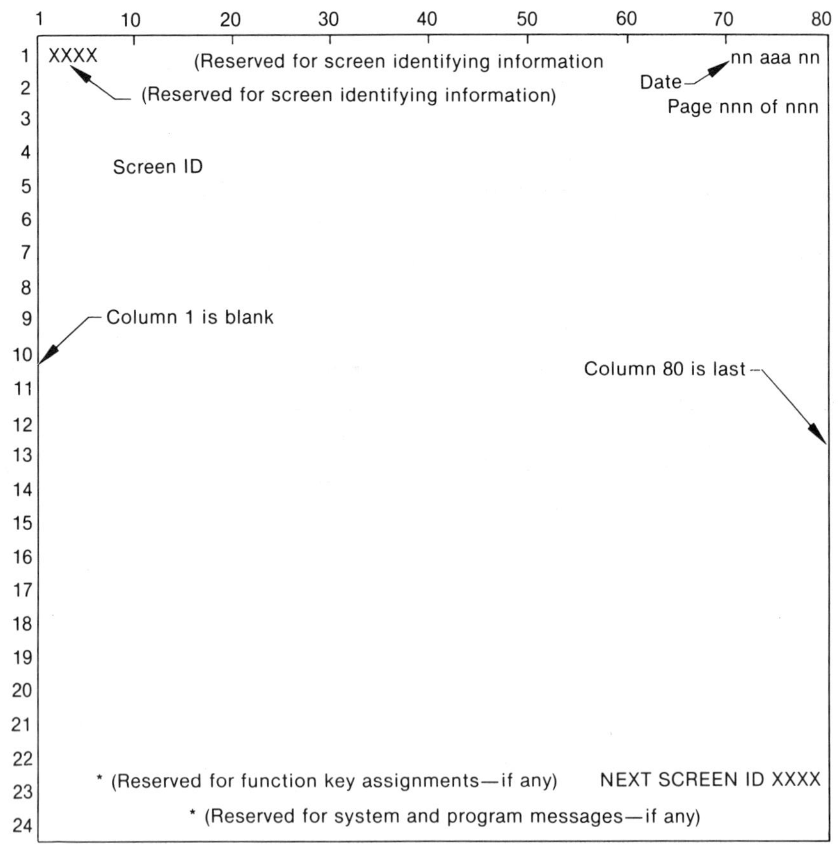

Figure 10. Display showing example of a standardized format (Source: Reference [18]).

Interframe Considerations – paging and scrolling

The "window to the world," as CRT screens in process control environments are often called, must be manipulated by the user. Unlike a traditional control board where all the instrumentation is immediately available, a CRT screen is restricted in the amount of information that can be presented at any particular time. As a result, the problem of how information can be accessed is an important issue. This section will address the various techniques for controlling and accessing information flow through the window. Brown et al. [18] refer to these elements of user control as follows: "One of the most critical determinants of user satisfaction and acceptance of a computer system is the extent to which the user feels in control of an interactive session." Two of the most common techniques for achieving user control are scrolling and hierarchical paging. (Another technique windowing, is discussed in the next section.) These approaches, their advantages, and applications will be discussed under this variable.

	Guidelines	Research Support	Source
	Paging/Scrolling		
1.	Whenever possible, all data relevant to the user's current transaction should be included in one display page (or frame).[a]	L	150, 5, 135, 36
2.	When the requested data exceed the capacity of a single display frame, the user should be given easy means to move back and forth among relevant displays either by paging or scrolling.[b]	L	23, 135
3.	When a list of numbered items exceeds one display page and must be paged/scrolled for its continuation, items should be numbered continuously in relation to the first item in the first display and should indicate the present maximum location. For example, use Line 63 of 157, not Page 3, Line 8.	L	5, 135, 47

4.	When lists or tables are of variable length and may extend beyond the limits of a single display page, their continuation and ending should be explicitly noted on the display.[c] For example, incomplete lists might be marked "continued on next page," or simply "continued." Concluding lists might add a note "end of list."	L	135
5.	When display output contains more than one page, the notation "page x of y" should appear on each display.[d]	L	150, 135
6.	The parameters of roll/scroll functions should refer to the data being guidelined, *not* to the window.[e] For example, "roll up 5 lines" should mean that the top five lines of data would disappear and five new lines would appear at the bottom; the window through which the data is viewed remains fixed.	L	135, 47, 18
7.	When the user may be exposed to different systems adopting different usage, any reference to scroll functions should consistently use functional terms such as forward and back (or next and previous) to refer to movement within a displayed data set rather than words implying spatial orientation (e.g., up and down).[f]	L	135
8.	When using a menu system, the user at all times should have access to the main menu.[g]	L	18
9.	Displays should indicate how to continue.[h]	L	18
10.	User-terminal interaction tasks that are repetitive, time-consuming, or complex should be assigned dedicated functions.[i]	L	18
11.	Required or frequently used data elements should be included on the earliest screens in the application transaction.	L	53
12.	Page design and content planning should minimize requirements for operator memory.	L	150, 5

Frame Hierarchy and Elements

13.	When pages are organized in a hierarchical fashion containing a number of different paths through the series, a visual audit trail of the choices should be available upon operator request.	L	150, 5
14.	Sectional coordinates should be used when large schematics must be panned or magnified.	L	150
15.	If the message is a variable option list, common elements should maintain their physical relationship to other recurring elements.	L	150
16.	A message should be available that provides explicit information to the user on how to move from one frame to another or how to select a different display.	L	5
17.	When the operator must step through multiple display levels, priority access should be provided to the more critical display levels.	L	146
18.	When the operator must step through multiple display levels, he or she should be provided with information identifying the current position within the sequence of levels.	L	146
19.	A similar display format should be used at each level of a multiple-level display.	L	146
20.	When the operator is required to accurately comprehend previously learned items appearing with a new list, the list should be kept small (about four to six items).	L	47
21.	Frequently appearing/disappearing commands/ subcommands should be placed in the same place on the screen.	L	47

[a] Do not rely on the user to remember data accurately from one display to the next [135].

[b] Paging/scrolling is acceptable when the user is looking for a specific data item but not when the user must discern some relationship among separately displayed sets of data [135].

[c] Short lists whose conclusion is evident from the display format need not be annotated in this way [135].

[d] A recommended format is to put this note immediately to the right of the display title. With such a consistent location, the page note might be displayed in dimmer characters. Leading zeros should not be used in the display of page numbers [135].

[e] Reference [135] also states that when a windowing orientation is maintained consistently, the wording of scroll functions should refer to the display page (or window) and not to the displayed data. In that case, the command "Up 10" would mean that ten lines of data will disappear from the bottom of the display and ten earlier lines will appear at the top.

[f] In that event, control of scroll functions should be implemented by keys marked with arrows, avoiding verbal labels, altogether [135].

[g] The user should not have to backtrack to return to the starting level in a hierarchical menu system. This capability can be provided by dedicating a program function key, touch field, or a cursor entry field to display MAIN MENU [18].

[h] Indicate on each screen the user response that is necessary to continue the interaction sequence. Do not leave the user viewing a screen with no indication of how to continue [18].

[i] User-terminal interaction tasks that are repetitive, time-consuming, or complex can be simplified by assigning dedicated functions (usually one user action) in system design. These actions should be accomplished by dedicated program function keys, dedicated light pen or touch fields, or other similar application. For example Function keys could be assigned to

- PAGE AHEAD
- HELP
- MAIN MENU
- PAGE BACK
- PRIOR LEVEL
- HARDCOPY, [18]

Comment. The above guidelines tend to reflect two underlying themes.

- When the operator must be working with several frames, attention should be devoted to reducing memory load and recall of information on previous screens.

- Some methods of providing a roadmap should be used in order to allow the operator to know where he or she is in the series.
- The merits of superiority between paging and scrolling have been given considerable attention in the literature. However, the issue of scrolling versus paging are, to some extent, task-dependent.

Interframe Considerations – Windowing

Windows exist on the user screen to provide a number of functions, the most popular of which is to allow for pull-down menus. Windows also allow users to integrate different tasks by working with several programs at the same time or by transferring information between applications [92]. An example of windowing is presented in Figure 11. Windows have become increasingly popular since the Apple Macintosh computer was first released in the early 1980's and now are popular in both the Gem and Microsoft windows product environments. Despite their advantages, there are some risks associated with windows regarding users becoming lost between applications or covering one application with another. Although there are few comprehensive guidelines available to the user, the following suggestions are offered.

	Guidelines	**Research Support**	**Source**
1.	The system should not allow for more than three applications to be run at a time.	L	Authors
2.	Once you quit a program (application), that window should close promptly.	L	Authors
3.	Windows should be consistent in their use of drop-down menus and/or icons.	L	Authors

4.	Navigation within the menus, whether mouse or direction keys are used, must be consistent.	L	Authors
5.	Windows should give the user feedback whenever he/she is in the process of combining applications.	L	Authors
6.	Dialogue boxes should be provided when necessary, to assist in defining menu options.	L	Authors
7.	Actions necessary for changing the size of a window should be consistent between windows.	L	Authors
8.	Active windows should be so labeled.	L	44
9.	The window label should be located at the top of the window border.	L	44

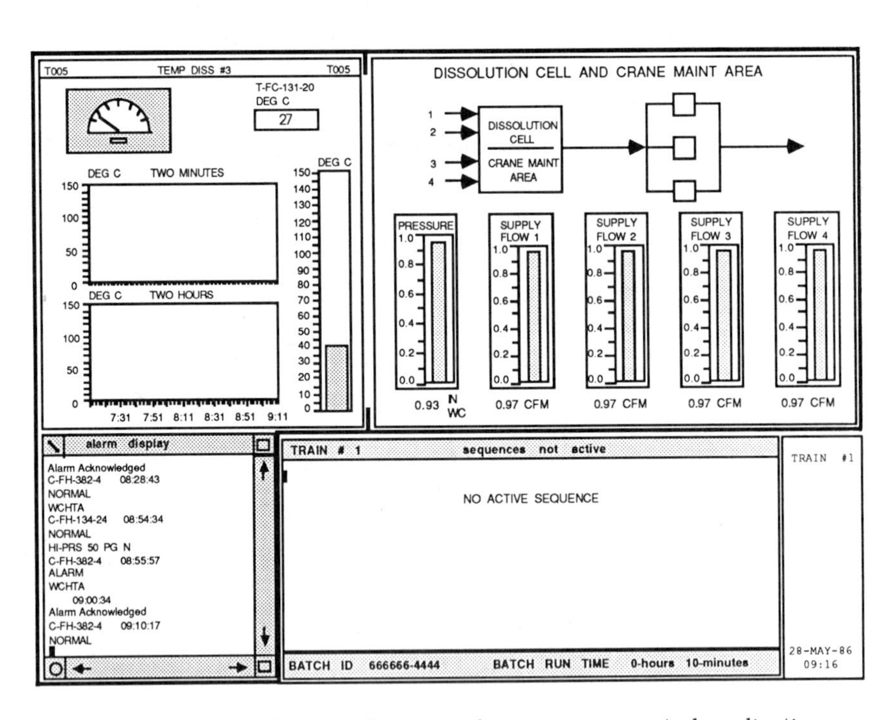

Figure 11. A windowing scheme used in a process control application.

10.	keyboard input should only affect the active window.	L	44
11.	Frequently appearing/disappearing commands/subcommands should be placed in the same location on the screen.	L	Authors

Visual Coding Dimensions

Visual coding of information, when applied effectively, can be a useful method for aiding operator performance. In many cases, coded information can ensure brevity and conserve space (i.e., pack information into a smaller space) without compromising interpretational quality. Conversely, poorly designed codes degrade performance. All coding designs should adhere to the principles of legibility, clarity, and consistency. Ambiguity and confusion should be minimized prior to implementing the code in the field. Within the context of graphic and alphanumeric codes, the sections which follow discuss these variables:

- Color
- Geometric shape
- Pictorial
- Magnitude

For purposes of this text, information highlighting is examined in a section entitled Enhancement Coding.

A set of general design guidelines that are applicable to all visual coding dimensions is provided in Table 3. This top-level table should supplement the guidelines and criteria presented for each of the following specific variables. It is recommended that Table 3 be reviewed prior to investigating the specific variables.

TABLE 3. GENERAL GUIDELINES FOR VISUAL CODING OF INFORMATION

- If a graphic display contains a high density of display elements, particularly overlapping and continuously updated features, then segregate the information using a coding or sequencing technique [81].
- Double coding dramatically increases user performance in information extraction. A redundant code should augment symbolic information, rather than add new symbols [81].
- Color is an extremely powerful coding technique for segregating information, but use it to assist the user at the primary or first level of interest. If a display is to be used for multiple tasks or if the system designer does not know what information is of prime importance to the user, color coding should be placed under user control [81].
- Techniques for reducing symbol density by segmenting displayed information should be standard functions on automated systems. Topographic information can be displayed sequentially, and an overlap between views facilitates user integration of information [81].
- Selective call-up seems to be an attractive solution to high-density, cluttered displays. However, penalties are incurred in the amount of time required to build a particular information display. Research is needed to examine means of effectively using this option [81].
- In selecting a code, consider the following:
    - Compatibility of the code with the kind of information that is to be coded,
    - Space required to use the code,
    - Association value of the code symbols,
    - Ease and accuracy with which the operator can understand the code,
    - Likelihood that the code will distract the operator or interfere with other codes, and

TABLE 3 (*continued*)

- Amount of information that needs to be coded and the amount that can be coded with each method [148].

- If combination codes are used, the respective meanings should be decodable separately without confusion. A maximum combination of two codes are recommended [160].
- Codes should be easy to learn and retain [160].
- Redundant cues (i.e., those which associate a symbol with a particular size) increase coding efficiency.[160].
- The principles of standardization and consistency of language and coding apply not only within a displayed screen but also within a module, an application program, and an entire information processing system. For a large corporation's information processing system, conventions for wording, coding, definitions, and procedures should be standardized at the company level. When users are forced to adapt to different conventions for each of several application programs they use, accuracy and efficiency suffer [18].
- Information coding should be used to discriminate among different classes of items presented simultaneously on the display screen [53].
- Meaningful codes should be used when possible. Codes should be clear and consistent with the user's expectations [18].
- If the type of information coding selected reduces legibility, is not distinct, or increases transmission time, it should not be used [114].
- Coding techniques shall be used to facilitate
  - Discrimination between individual displays,
  - Identification of functionally related displays,
  - Indication of relationships between displays, and
  - Identification of critical information within a display [146].

TABLE 3 (*continued*)

- Graphic code consistency–graphic codes, used separately or in combination–should have the same meaning in all applications [150].

Color

The use of colors as a coding dimension can increase an operator's information-gathering and processing capabilities. Specific colors can be used to:

- reduce confusion and minimize control search time,
- aid in location,
- provide necessary redundancy to other coding methods,
- identify emergency controls, and
- foster immediate recognition of the meaning or function of a control.

The use of colors in display design is well researched. The bulk of this research indicates colors to be an extremely effective dimension for search and identification tasks [28]. R. E. Christ noted that, compared to other coding dimensions such as geometric shapes and size, color was superior [28]. Other research efforts have been related to determining the number of distinct colors that persons with normal color vision can differentiate on an absolute basis [27], [31], [100].

The actual perception of colors is determined by the wavelength along the visible spectrum. In general, this visible spectrum ranges from 380 to 700 nm. The reds and oranges are derived from the higher wavelengths, while violets and blues are at the lower end of the spectrum. Within these boundaries of electromagnetic radiation, color is further defined as a function of hue, saturation, lightness, and brightness.

Hue can be considered the determinate of colors or, in a more perceptual context, the experience of colors which allows us to discriminate between otherwise identified objects. Hue is not used in terms of defining white or black or gray or in terms of describing a light source. Along the visible spectrum, hue can also be interpreted as the perceived colors having dominant wavelength components (e.g., the color is red because the dominant wavelength is approximately 700 nm).

Saturation refers to the purity of perceived colors. The more white light is mixed with a pure saturated hue, the more desaturated the resulting colors experience. Lightness is often analogous to saturation. This characteristic includes the gamut of achromatic colors ranging from white through gray to black. No trace of hue is present. The combination of various hues and achromatic colors produces a desaturated hue, with the level of saturation depending upon the relative amounts of each component.

Brightness is the dimension of perceptual experience corresponding roughly to light intensity. Alterations in brightness can significantly alter the perception of a given hue. For example, increases in brightness levels cause hue to increase in saturation. Eventually, at upper limits of brightness, hues can appear washed out and approach luminescent white.

All perceived colors on the visible spectrum are also produced by mixing the three primary colors for light: red (R), green (G), and blue (B). All colorgraphic CRTs have a colors gun for each of these primary colors. By specifying the relative amounts of each primary mixed together, any spectral color (C) can be indicated by an appropriate equation:

$$C = xR + yG + zB \tag{1}$$

where

x, y, z = proportions of each wavelength.

The following guidelines address various criteria for assessing the effectiveness of colors as it is implemented in the coding of CRT displays, even though a sizable portion of these guidelines was developed from surface colors. In addition, specific recommendations for candidate color sets are provided.

Guidelines	**Research Support**	**Source**
General		
1. Colors should be used as a formatting aid to assist in structuring a screen and as a code to categorize information or data.	L	7, 135, 53, 18, 46, 148
2. Color coding should not create unplanned or obvious new patterns on the screen.[a]	L	7
3. Color coding should be applied as an additional aid to the user on displays that have already been formatted as effectively as possible on a single color.[b]	L	135, 18
4. When color coding is used, it should be redundant with some other feature in data display, such as symbology.[c]	L	7, 5, 135
User Control		
5. The unit should, as a minimum, be provided with the following controls:	L	7

a. A foreground intensity control separate from the background intensity control.

b. A capability for making grid lines half as intense as the rest of the display.

c. Enough intensity control variable to accommodate very low ambient illumination and the higher levels normally found in office work areas (5 to 150 fc).

Color Consistency

6.	Color meanings should be consistent with traditional color expectancies.[d]	Y	150, 135, 53
7.	Color coding should be consistent within a frame, from frame to frame, and with other color-coded systems in the control room (see Table 4).[e]	Y	150, 5 53, 18

TABLE 4. ASSOCIATIONS AND RELATED CHARACTERISTICS FOR COLORS TYPICALLY USED IN PANEL DESIGN AND RELATED APPLICATIONS IN NUCLEAR POWER PLANTS (From Reference [46])

Color	Associated Meanings	Attention-Getting Value	Contrasts Well With
Red[a]	unsafe danger alarm state open/flowing closed/stopped	good	white
Yellow	hazard caution abnormal state oil	good	black dark blue
Green[a]	safe satisfactory normal state open/flowing closed/stopped	poor	white

[a] Meanings associated with red and green colors differ, depending on past experience. Personnel with previous fossil-fuel plant experience typically associate an open/flowing state with red and a closed/stop state with green, but reverse associations typically exist for personnel with previous Navy experience

TABLE 4 (*continued*)

Light blue (cyan)	advisory aerated water cool	poor	black
Dark blue	advisory untreated water	poor	white
Magenta	alarm state	good	white
White	advisory steam	poor	green black red dark blue magenta
Black	background	poor	white light blue yellow

8.	Color codes should conform to color meanings that already exist in the user's job.[f]	L	146, 53 46
9.	Specific color selections should conform to the general guidelines of Table 5 and to the following specific recommendations:		
	a. The most generally used colors should be red, green, yellow, and blue. Other acceptable colors are orange, yellow-green, blue-green, and violet.	L	53
	b. When blue headings, numbers, or alphabetic characters are used, the background should be black.	Y	7
	c. Yellow should not be used on a white background because of the very low contrast.	Y	7
	d. Yellow should not be used on a green background due to a vibrating effect to the eye.	Y	7

TABLE 5. GENERAL GUIDELINES FOR COLOR SELECTION (Source: Reference 150)

Red - Good attention-getting color. Associated with danger.

Yellow – (amber)--Good attention-getting color. Associated with caution.

Green – A nonattention-getting color. Associated with satisfactory conditions.

Black – Normally used as the background color, i.e., the color of blank character spaces. Also used as the action character when reverse field coding is employed.

White – A nonattention-getting color. It should be used for standard alphanumeric text or tables where the information is contained in the characters and not the color. Might also be used for labels, coordinate axes, dividing lines, demarcation brackets, etc.

Cyan (light blue) - (Same as white)–Might be used in conjunction with white to provide some amount of noncritical discrimination (e.g., use cyan for tabular column headings and demarcation lines; use white for alphanumeric data).

Blue(dark) – Poor contrast with dark background. Not recommended for attention-getting purposes or for information-bearing data. Use for labels, and other advisory type messages.

Magenta – A harsh color to the eye. Should be used sparingly and for attention-getting purposes.

Orange – Good attention-getting color. Care must be taken that hue is selected to be readily differentiable from red, yellow, and white.

---

e. White should be used for very important information.[g]	L	7
f. The selected colors should yield satisfactory color contrast for color–deficient users.	L	46

Guidelines	Research Support	Source
g. The user should be able to discriminate the selected color on an absolute basis (see Table 6).[h]	L	53, 46
h. Selected colors should be usable in all control room applications (e.g., panel surfaces, labels, CRTs, indicator light bulbs or filters, console surfaces).		46
i. Blue should be used only for background features in a display, not for critical data.	L	135
j. Whenever possible, red and green should not be used in combination. Use of red symbols/characters on a green background especially should be avoided.	L	150
10. If a pattern of color is intended to display a function, the selected colors should indicate the state of the system.	L	7

TABLE 6. SPECTRAL COLORS IDENTIFIED ACCURATELY WITH LITTLE TRAINING (Source: Reference [153])

642 nm	Red	515 nm	Green
610 nm	Orange	504 nm	Green-blue
596 nm	Orange-yellow	494 nm	Blue-green
582 nm	Yellow	476 nm	Blue
556 nm	Yellow-green	430 nm	Violet

11. Colors with high contrast should be selected for parameters and features that must "catch" the operator's attention.	L	7
12. In general, backgrounds should not be brighter than foregrounds.	L	7
13. Extreme color contrasts should be avoided.[i]	Y	7
14. Colors should be specified as a precise wavelength rather than a hue (red, green, violet, etc.) (see Table 6).	L	7
15. If difference in brightness (intensity) is used as a coding mechanism, perceived brightness should be used rather than absolute brightness.[j]	L	7, 18
16. Each color should represent only <u>one</u> category of displayed data.	L	150, 135, 18, 46
<u>Number of Colors</u>		
17. If color discrimination is required, do not use more than eight colors. Alphanumeric screens should display no more than four colors at one time.[k]	L	135, 53
<u>Control room Lighting</u>		
18. Colored ambient lighting should not be used in conjunction with color-coded CRTs.	L	7
19. If ambient illumination cannot be controlled, hoods should be provided that block out light and glare.	L	7
20. High-pressure sodium should not be used as an ambient-light medium for CRT viewing.	L	7
<u>Maintenance of Colors</u>		
21. Color displays should be periodically adjusted to maintain proper registration of images.[l]	L	18

Color Applications for Mimics

22. Identify potential applications of color in mimic designs. Mimics can incorporate color to differentiate process flow paths. For example, blue may be used to code water lines; white, steam lines; yellow, oil lines; and so forth. Such color differential is potentially valuable in helping operators to sort out complex interrelationships. — Y — 53

Color for Highlighting

23. Color is extremely effective for highlighting related data that are spread around on a display, such as data of a particular status or category. Color may effectively aid in the location of headings, out-of-tolerance data, newly entered data, data requiring attention, etc. Search for data on a display is facilitated by color coding if the color of the data sought is known. — Y — 18

24. A monochromatic format can be used if the flash is twice as intense as the rest of the display. This will draw the operator's attention as effectively as if color had been added. Fewer colors can be used if they can be adjusted for intensity. — Y — 7

Misuse of Color

25. Do not overuse color. Use of too many colors may make a screen confusing or unpleasant to look at.[m] — Y — 53

26. Color capability must be used conservatively in the design of display screens. Arbitrary use of multiple colors may cause the screen to appear busy or cluttered, and it may reduce the likelihood that the information in color codes on that screen or on other screens will be interpreted appropriately and quickly. In general, color should be added to the base color display only if it will help the user in performing a task. — N — 18

27.	Do not use color coding in an attempt to compensate for poor display format; redesign the display instead.	L	135
28.	The use of color should not be distracting to the user.	L	53

---

[a] Patterns of color on the CRT face can either impede or enhance operator performance. Test each format for distracting visual noise before finalizing the color-code assignments [7].

[b] Reference [135] states, "Displayed data should provide necessary information even when viewed at a monochromatic display terminal, or hardcopy printout, or when viewed by a user with defective color vision."
Reference [18] states, "In designing a screen, color is applied as an additional aid to the user, though generally the format should be such that it is effective when displayed in a single color."

[c] Reference [18] states, "Information must not be conveyed solely by color change if the information may be accessed from monochromatic as well as color terminals and when color may be lost in printed (hardcopy) versions. Where both kinds of terminals may be in use, color must be limited to assisting the user of the color terminal by highlighting, aiding in categorization, or clarifying the relatedness of data, without sacrificing important information to the user of the monochromatic display. Attention should also be addressed to the sometimes misleading appearance of color in photographs of display screens which may be intended for oral presentations. Printed output will not show color or flashing and probably will not indicate highlighting unless a color printer is used."
Reference [5] states, "Keep in mind when using color coding in displays that some colors are not discriminative by some users. Some users will have defective color vision."

[d] Reference [150] states, "The many hues (colors) and saturations should, where applicable, equate with the commonly understood meaning of those colors. The following specific meanings for selected colors should apply when these colors are used in CRT displays:

- Red–Unsafe condition, danger, immediate operator action;required, or critical parameter value out of tolerance.

- Green–Safe condition, no operator action required, or parameter value is within tolerance.

- Yellow/Amber–Hazard, potentially unsafe, caution, attention required, marginal parameter value exists."

[e] Reference [5] states, "To illustrate the extent of today's use of inconsistent color coding schemes, a study of color coding of displays in nuclear power plant control rooms found that red denoted 'on' or 'flow' and green denoted 'off' or 'no flow' In addition, the investigators found that this use of red-green coding had been mingled with military and other color-coding schemes to the point where almost every designer used their own preferred scheme. To avoid this kind of confusion, designers should rely on standard color-coding schemes."

[f] Reference [53] states, "Color meanings will be more easily learned if color codes conform to color meanings that already exist in a person's job. Color codes employing different meanings will be much more difficult to use." Reference [46] states, "Determine appropriate associations for the selected colors." Table 4 indicates some associations and related characteristics for colors typically used in a variety of nuclear power plant applications. Reference [53] further states, "Where previous associations differ among personnel due to differences in experiences [e.g., personnel with Navy versus fossil-fuel plant experience], select colors that are not in conflict with prior associations and recommend that adequate opportunities be provided for practicing the selected color associations."

Reference [146] recommends the following color meanings:

Color	State	Result
Flashing red	Emergency	Immediate user action
Red	Alert	Corrective/override action must be taken
Yellow	Advise	Caution; recheck is necessary
Green	Proceed	Condition satisfactory
White	Normal Transitory function	No right or wrong indication

[g] Use the white color for very important information so that if one of the color guns fails, the information will not be lost from the screen. (White uses all three color guns in the CRT.) [7].

[h] Reference [53] observes that the number of easily identifiable spectral colors depends on luminance, size (in visual angle), and color of the lights. The ten spectral colors in Table 6 can be identified with a 2% error after a relatively short training period when the lights have a luminance of 1 mL or

more. The angular substance of the color source should not be less then 15 minutes of arc for highly accurate color recognition.

[i] Extreme color contrasts can form complementary afterimages due to rod receptor fatigue and should be avoided. For example, after staring at a yellow target for more than 10 seconds on a black background, the eye will perceive a green afterimage when it is suddenly averted [7].

[j] Reference [18] states, "The human eye is not equally sensitive to all colors, nor are its optics color-corrected. The eye is very much more sensitive to colors near the center of the visible colors spectrum (orange, yellow, and green) than it is to colors at the extremes of the spectrum, such as red or blue. This variation in sensitivity causes colors at the center of the spectrum to appear much brighter than colors having equal energy at the extremes of the spectrum."

[k] Reference [135] states, "Perhaps as many as 11 different colors might be reliably distinguished, or even more for trained observers, but as a practical matter it will prove safer to use no more than five or six."

[l] Reference [18] states, "Color displays require periodic adjustment to maintain proper registration of images. When out of adjustment, characters formed by a combination of primary colors (pink, yellow, turquoise, and white), may appear as characters in each of the component primary colors. Pink characters, for example, may appear as red characters with blue shadows or echoes on a display which is out of adjustment. Novice users should be made aware of this and the adjustment procedure early in their training."

[m] Reference [53] states, "Color's high attention-getting quality may be distracting, causing a person to

- Notice differences in colors, regardless of whether they have task-related meaning and

- Visually group items of the same colors in a way that is unrelated to the task or in conflict with another task-related group of items."

Comment. For purposes of evaluating process CRT displays, the guidelines presented in References [7], [150], and [46] are perhaps the most relevant. Some controversy between the meanings of particular colors codes was discussed by Reference [46]. The authors suggest that adherence to Reference [146] and general population stereotypes should take precedence over alternative conventions.

However, as with most coding methods, some mode of standardization and consistency should be observed throughout the control room. If a previous standard is already in place and consistent, then recommendations for altering existing colors meanings could be devastating. If the colors meanings are already standardized, the screen should be examined for number of codes implemented and whether or not they can be easily discriminated by the operators.

Geometric Shape

Dreyfus [40] considered graphic symbols to be one of three types: (a) representational–accurate simplified pictures of objects, e.g., skull and crossbones; (b) abstract–graphic symbols retaining only a faint resemblance to the original concept, e.g., signs of the zodiac; and (c) arbitrary–symbols that have been invented and need to be learned, e.g., inverted triangle indicating "yield." In actual practice, many of these graphical codes may not be solely confined into any one category. For example, a roadway sign in the shape of a pentagon (arbitrary portion) may have a silhouette of two children walking (representational portion) to indicate a school crossing.

This section will be devoted primarily to the arbitrary category of graphical symbols. This includes the geometric forms and shapes which do not have a direct reference to the real world and also require some level of formal training before they can be learned and implemented. Pictorial graphical coding (often referred to as iconic coding) will be addressed in another section.

Guidelines	Research Support	Source
1. Geometric shapes should be considered for discriminating different categories of data on graphic displays.	L	135
2. Alphabets of geometric shapes should be limited to a maximum of 15 different symbols.[a]	Y	67, 53, 135, 10, 133

<u>Design of Shapes</u>

3.	When geometric shape (symbol) coding is used, the basic symbols should vary widely in shape.[b]	Y	150, 138, 38
4.	Symbols should subtend a minimum of 20 min of arc. If the viewing distance is longer than the normal 28 in., it should form a visual angle of about 22 min of arc.[c]	Y	13
5.	The stroke width-to-height ratio should be 1:8 or 1:10 for symbols of 0.4 in. or larger viewed up to a distance of 7 ft.	Y	13

<u>Redundancy</u>

6.	When efficiency of decoding is important, redundant cues (such as size difference) should be used.[d]	Y	13
7.	When rate of comprehension and detection is important, graphical coding should be used rather than word messages.[e]	Y	79

<u>Standardization</u>

8.	The assignment of shape codes should be consistent for all displays and should be based upon an established standard.[f]	Y	135, 97

---

[a] If [an] alphabet [of 15 symbols] is too small, it may be possible to use component shapes in combination, as in some military symbol codes [135]. The number of identifiable shapes in letter and number form is unlimited. For shapes in geometric form, however, learning and retention are limited to around fifteen (five recommended) unless special training and continued practice are undertaken [53], [10].
Reference [67] states, "Sleight [133] found 15 geometric shapes that are highly discriminative if their apparent size is at least 12 min of visual angle and there is high contrast with the background. These included common shapes such as the triangle, circle, star, square, and pentagon. These shapes may be used in coding, but their meaning must be taught. Other shapes among the 15, such as the swastika, heart, half-moon, or aircraft silhouette, have certain conventional meanings that could detract from their usage in other applications."

[b] Reference [138] states that search time and errors increase when highly similar shapes are used. Reference [38] states that variations of a single form such as an ellipse are not desirable. References [66] and [64] find that the "triangle" is the most discriminative geometric shape when compared with other forms (e.g., star, rectangle, circle, square, parallelogram).

[c] The size of any symbol can be measured at the surface of the screen, but this does not determine the visibility of the symbol. What is important is the angle that is subtended by the symbol at the viewer's eye, and this angle depends on symbol height and viewing distance [45].
The angle which any object subtends can be calculated by the equation

$$B = (S \times 360 \times 60)/2\pi D \tag{2}$$

where

S = size of the image on the screen

π = 3.14

D = distance to the screen (same units as S)

B = arc which the object subtends (minutes).

For the designer, it is probably more useful to rearrange this equation into the form

$$S = 2\pi DB/(360 \times 60). \tag{3}$$

[d] Reference [88] states that under normal viewing conditions, such as white light and daylight, people use area and jaggedness most frequently to describe shapes.
Reference [26] states that, under poor viewing conditions such as subdued light, glare, or fading, area and the largest dimensions are used to distinguish different shapes.

e. In general, symbolic/pictographic signs are more quickly detected and comprehended than word messages.

[f] Although shape codes can often be mnemonic in form, their interpretation will generally rely on learned association as well as immediate perception. Existing user standards must be taken into account by the display designer.

Comment. Research evidence indicates a superiority of pictographics over geometric or shape coding [79]. Van Cott and Kinkade [153] stated, "Pictorial shapes depicting the real-world objects they represent, such as aircraft, ship, and missile silhouettes, can be easily learned, remembered, and used." This observation is supported by Smith and Thomas, [138] who found that pictorial shapes of military objects are superior to highly discriminative geometric forms. However, pictorial shapes must not be similar in form. Otherwise search time and errors increase.

As indicated by the above guidelines, the 15 discriminative shapes identified by Sleight [133] are most often recommended if a geometric coding methodology is implemented. With minor modification, such as removal of those shapes with conventional meanings (swastika, airplane, heart, etc.), a limited coding set could be derived for process control application. The earlier evidence also supports the use of the triangle as being the most discriminative; therefore, it should be included as a prime candidate symbol for the limited set. If these shapes (identified by Sleight) are to be used for indicators of warning, the following data from traffic safety signs might be applicable. In that study, Riley et al. [124] evaluated various sign shapes to determine those that were preferred indicators of warning. Stimuli consisted of nine standard shapes and an additional 10 shapes based upon their similarity to the standards.

The analysis resulted in a ranking of the most preferred shapes. Based on that study, Riley established the following conclusions:

- Shapes that appear unstable tend to be more preferred as warnings.
- Since shapes of signs and labels, do have a relationship to warning indication, the shape hierarchy determined from this study should be considered in guiding the design of warning indicators.

In addition, care should be taken to ensure that the selected shapes are of sufficient size to be easily identified from a predetermined viewing distance. Symbol size can be quantitatively determined using the equation described in footnote c in the previous guidelines [45].

Pictorial

This coding strategy contrasts sharply with simple geometric or shape coding. Pictorial codes are essentially pictorial representations of the objects and events they reproduce by various laws of projection. According to Dreyfus [40], geometric or shape coding would be categorized as arbitrary. Pictorial coding comprises the category of representational symbols and, in some cases, abstract symbols as defined by Dreyfus. A more detailed definition of pictorial coding has been defined by Schiff [125] as "drawn, painted, etched, or projected structure, usually representing a scene or object. It varies from photographic realistic pictures to representational paintings, mimics, or drawings to nonrepresentational sketches, line drawings, or abstract designs." In the literature, pictorial coding is often referred to as iconic coding. In general, less training is needed for identification of referents through pictorial coding. If pictograms are designed in accordance with expectations of the user population, the users already know the relationships. This section is devoted to the appropriate use of pictures as a coding method.

	Guidelines	**Research Support**	**Source**
1.	Pictographs should have obvious meanings, and the meaning should be tested in the user population.	Y	67, 46, 29
2.	Symbols should be consistently applied.	L	46, 97
	Mixing Pictographs and Words		
3.	Words and symbols should not be used alternately.[a]	Y	160, 46
4.	Symbols should be used to represent equipment components and process flow or signal paths, along with numerical or coded data reflecting inputs and outputs associated with equipment.	L	46

Design of Pictographs

5.	Symbols used to represent equipment components vary widely in shape and should be similar to those used in piping and instrumentation drawings.	L	46
6.	About six different symbols, with 20 being an upper limit, should be used.[b]	Y	10, 46
7.	When iconic symbols are used, solid forms without unnecessary detail are preferred.[c]	Y	91, 80
8.	The visual saliency of those features that must remain redundant across members of a symbol set should be minimized.	Y	29
9.	A closed figure enhances the perceptual process and should be used unless there is reason for the outline to be discontinuous.	Y	91

Labeling Conventions

10.	When letters are used, perhaps to annotate geometric display symbols, lowercase letters should be used to improve discriminability.	Y	29
11.	For search and identification tasks or whenever there is any doubt as to whether some observers will be able to understand the pictorial, both pictorial and word labels, should be used.	Y	160, 36, 46

Redundancy

12.	Given that a sufficient number of dimensions are available to portray required information parameters, multidimensional display codes should be used.	Y	29

Distortion

13.	To minimize distortion (especially under degraded viewing conditions), well-learned or unitized symbol designs should be used.	Y	29

	Pictorial Simplicity		
14.	The symbols should be as simple as possible, consistent with the inclusion of features that are necessary.[d]	Y	91, 160
	Control room Lighting		
15.	The pictorial pattern should be identifiable from the maximum viewing distance and/or under minimal ambient lighting conditions.[e]	Y	160
	Orientation		
16.	Pictorial symbols should always be oriented "upright."[f]	Y	160
	Iconic Representation		
17.	Icons should not be used when display resolution is low.	L	86
18.	A label should be associated with each icon.	L	86
19.	Abstracts (icons) may be of either a literal, functional, or operations type.	L	86
20.	To the extent possible, icons should concur with existing population or industry stereotypes.	L	86

---

[a] Alternate use of symbols and words could cause confusion and retard task performance [46]. If a particular situation includes items that cannot be completely pictorialized, i.e., if some items are easy to pictorialize but others are not, it is better to stick to a word labeling system; observers will only be confused if they find some pictorials and some word labels, on an operator panel [160].

[b] For pictorial shapes, Reference [46] recommends that the maximum number of code steps be 30, with 10 recommended. The maximum number assumes a high training and use level of the code. Also, a five percent error in decoding must be expected. The recommended number assumes operational conditions and a need for high accuracy.

[c] Reference [45] recommends as a method of indicating equipment status using a hollow (outline) element to indicate the element is off, closed, or

showing no flow and using a solid element to indicate that it is on, open, or showing flow.
Reference [160] recommends that some type of border should always be used around a pictorial; otherwise, it may blend with background images.

[d] Reference [160] states, "In creating a pictorial to represent some actual visual element, provide just enough detail to make the symbol recognizable and no more. Fine detail often cannot be seen under certain distance and lighting conditions and may serve only to distort the impression the symbol creates. On the other hand, be extremely careful not to overstylize pictorial symbols just to create an artistic rendition. Such stylization often makes all symbols begin to look alike."

[e] Reference [160] states, "Some pictorial patterns may be effective only when the viewing distance and lighting conditions are optimum; be sure that a particular pictorial pattern does not lose its identity when it becomes smaller or more distant and/or when the ambient lighting and/or atmospheric conditions are not good."

[f] Reference [160] states, "Do not place pictorials on components that may be reoriented, i.e., turned or moved so that the pictorial may not appear right side up. Although the observer might figure out what the pictorial is after considerable study, a prime purpose of the single pictorial is to elicit a quick response with a single symbol, as opposed to several words in a label."

Comment. The majority of the research data on pictorial coding schemes was conducted specifically for military or roadway environment applications. Very little information is available for specific process control systems. An example of a unique pictorial coding scheme used to indicate a set of key plant parameters for nuclear power operations is illustrated in Figure 12. The most pervasive guidelines in this area have been adapted primarily from the engineering and drafting disciplines. The basic source for these guidelines is documented in ANSI Y32.11 [59]. These ANSI voluntary standards are often seen in piping and instrumentation diagrams (P&IDs). The meanings of some of these forms are shown in Figure 13. The Instrument Society of America (ISA) [76] has also prepared a standard for graphic displays which will be used for process control displays in a variety of industrial applications. These standards are loosely based on ANSI Y32.11. A word of caution is advisable before acceptance of these P&ID conventions to process control environments. Experimental data are limited, and problems

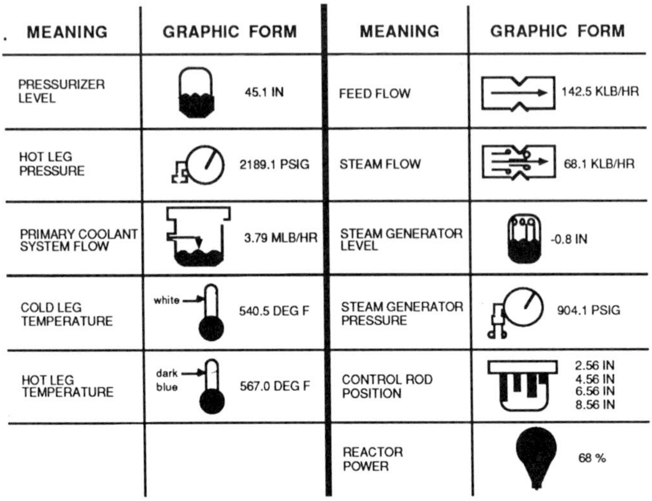

MEANING	GRAPHIC FORM	MEANING	GRAPHIC FORM
PRESSURIZER LEVEL	45.1 IN	FEED FLOW	142.5 KLB/HR
HOT LEG PRESSURE	2189.1 PSIG	STEAM FLOW	68.1 KLB/HR
PRIMARY COOLANT SYSTEM FLOW	3.79 MLB/HR	STEAM GENERATOR LEVEL	-0.8 IN
COLD LEG TEMPERATURE	white 540.5 DEG F	STEAM GENERATOR PRESSURE	904.1 PSIG
HOT LEG TEMPERATURE	dark blue 567.0 DEG F	CONTROL ROD POSITION	2.56 IN 4.56 IN 6.56 IN 8.56 IN
		REACTOR POWER	68 %

Figure 12. Pictorial codes used to indicate key plant parameters in nuclear power operations.

may arise when generalizing these standards to plant process control. True, the ANSI standards are useful building blocks, but further research and development are needed before specific recommendations can be addressed.

The above guidelines can best be summarized by adherence to the general recommendations shown in Table 7.

Magnitude

There are occasions when it might be feasible to select a common geometric shape (circle, square, triangle, etc.) and create an alphabet by varying some of its characteristics (e.g., changing the height-to-width ratio of a rectangle). Techniques such as these generally fall into the domain of magnitude coding. Magnitude coding can also include dimensions such as brightness and flash rate. This section is limited to coding techniques applicable to area and length. Alternative magnitude coding dimensions are discussed under the

Meaning	Graphic Form
Positive-displacement pump	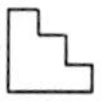
Centrifugal pump	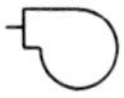
Steam generator	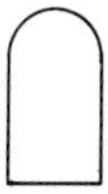
Tank	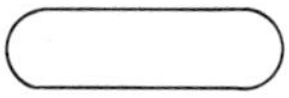
Valve	
Reactor vessel	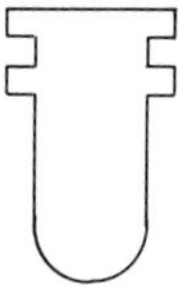

Figure 13. Pictographic codes which conform to conventional piping and instrumentation diagrams (P&IDs).

subcategory Enhancement Coding. Experimentally derived symbol alphabets, general guidelines, and techniques for developing these alphabets will be discussed in this section.

Guidelines	Research Support	Source
1. When symbol size is used for coding, the intermediate symbols should be spaced logarithmically between the two extremes (largest and smallest).[a]	Y	153

TABLE 7. OVERVIEW OF GENERAL GUIDELINES FOR PICTORIAL CODING

1. The apparent meanings of all candidate pictorial codes should be tested for a sample of the user population.
2. The coding symbols selected should be consistent for all applications. For example, the use of symbols in CRT displays should be coordinated with those used on all panels throughout the control room.
3. Alternate use of symbols and word labels, should be avoided.
4. The implementations of P&ID (ANSI) graphic symbols should not be accepted at face value. The population should be tested regardless of the standards adapted.
5. Use letters when necessary to annotate pictorial codes which are not readily obvious.
6. All graphical codes should be closed, unitized, simple, produce a high contrast with the background, and of sufficient size to be interpreted at a predetermined viewing distance (apply the technique for calculating symbol size in the section for shape coding if this dimension is questionable).

2.	When the symbol size is to be proportional to the data value, the scaled parameter should be the symbol area rather than a linear dimension such as diameter.[b]	Y	98
3.	For area coding, the maximum number of codesteps should be six, with three recommended.[c]	Y	135, 10, 4
4.	For length coding, the maximum number of code steps should be six, with three recommended.[d].	Y	153, 135, 10
5.	Size coding should be used only when displays are not crowded.	Y	135, 10

6. When size coding is used, a larger symbol should be at least 1.5 times the height of the next smaller symbol. Y 135

---

[a] To logarithmically space or scale the steps to get the same amount of accuracy all along the scale, apply the following general rule:

1. Identify the upper and lower limits of the scale.

2. Identify the number of steps.

3. To obtain the required order of magnitude between each step, apply the following formula:

$$R^{(S-1)} = A_{max}/A_{min} \tag{4}$$

where

R = required order of magnitude

S = number of steps

A_{max} = upper limit of the scale

A_{min} = lower limit of the scale.

For example, suppose an area coding alphabet of five circles is desired. Given

$A_{max} = 1.0 \text{ in}^2$

$A_{min} = 0.01 \text{ in}^2$

$S = 5$

Then using the equation

$R^{(S-1)} = A_{max}/A_{min} = 1.0 \text{ in}^2/0.01 \text{ in}^2$

$R^4 = 100$

$R = (100)^{1/4} = 100^{0.25}$

$R = 3.2.$

The required areas for a symbol alphabet of S circles would be 0.01, 0.032, 0.10, 0.32, and 1.0 in^2, with each area being 3.2 times the next smaller one [153].

[b] The sizes of graduated circles are generally scaled with the areas of the circles made proportional to the data values to be represented. If diameters are used for scaling the information, serious graphic problems arise [98]. Research evidence indicates if the areas of the circles are made proportional to the data values, differences in the sizes will be underestimated. These findings are not restricted to circles. It also applies to squares, triangles, and a multitude of other two-dimensional symbols. In summary, graphic symbols are not effective when scaled in a strict mathematical relationship to a set of data. However, the use of training and user access to an accurate detailed legend on the display may minimize the error involved in reading these types of symbols [98].

[c] The maximum numbers assume a high training and use level of the code. Also, five percent error in decoding must be expected. The recommended number assumes operational conditions and a need for high accuracy [10]. References [135] and [4] recommend five as the maximum number of symbols.

[d] The maximum number assumes a high training and use level of the code. Also, five percent error in decoding must be expected. The recommended number assumes operational conditions and a need for high accuracy [10]. Reference [135] recommends a maximum of four steps for reliable decoding. Reference [53] recommends a maximum of five steps.

Comment. Van Cott and Kinkade [153] stated, "This type of coding actually combines variations of shape, area, or size and would be particularly suitable for CRT display where dynamic variation in shape and area can be used as combined codes." The utility of an area/length coding method by systematically varying some parameter of a basic geometric form has merit if it adheres to the following restrictions:

- Values for a selected alphabet must be equally spaced on a logarithmic scale. Direct quantitative correlations of symbol size with a parameter value are not effective.
- The symbol alphabet must be small (usually less than five) to ensure accuracy.
- Large symbol space is required.

In addition, care should be taken to ensure that the CRT screen and upper and lower boundaries are sufficient for maximum discrimination at the prescribed viewing distances and resolution.

The factor for size coding of 1.5 times the height of the next smaller symbol was originally referenced in MIL-STD 1472C [146] and cited by Smith and Mosier [135] in the above guidelines. In the authors' opinion, this may be a reasonable hard and fast rule of thumb. However, a more precise logarithmic scaling approach (as prescribed by Van Cott and Kinkade [153]) on a case-by-case basis is recommended.

Enhancement Coding Dimensions

The previous category (Visual Coding Dimensions) examined alternative methods for presentation of information to the operator using a variety of coding dimensions. This category continues review of coding dimensions with a more specialized function i.e., those variables that accent critical information to the viewer. Alternatively, effective enhancement coding techniques will make the information more prominent to help attract the attention of the operator to specific items of information in the display. On a more general level, enhancement coding aids the operator to visually distinguish between classes or categories of information. Cakir et al. [23] reported, "The provision of one or more enhancement coding dimensions is usually an essential attribute of the VDTs (Visual Display Terminals) in an interactive system in which the information on the display screen is to be searched and manipulated." The majority of the guidelines reviewed often refers to enhancement coding as highlighting. For example, Engel and Granda [47] define highlighting as "the art of emphasizing something (e.g., data, field, title, or message) on the display screen." Cakir et al. [23] define enhancement coding as "the use of graphical techniques...to enhance the visual appearance of selected parts of the information on the display screen." Even though there may be subtle differences, these terms are considered synonymous (and are often used interchangeably) by the authors throughout the discussion of this document.

Highlighting and enhancement coding techniques rely on several coding dimensions. This subsection examines those variables having the most relevance to applications in process control systems:

- Brightness
- Blink
- Image reversal
- Auditory
- Voice
- Audio-visual
- Other techniques

A set of general guidelines which are applicable to all enhancement coding dimensions is shown in Table 8. This top-level table should supplement the guidelines and criteria presented for each specific dimension that follows. Those guidelines should be used as supporting information with each of the specific variables. It is recommended that those guidelines be reviewed prior to investigating the specific variables.

TABLE 8. GENERAL GUIDELINES FOR ENHANCEMENT CODING OF INFORMATION

- Highlighting methods which have information value beyond their attention-getting quality should have the same meaning in all applications [150].
- Recommended methods of highlighting should be used to attract attention to any displayed data that require rapid response [46].
- Highlighting can be a means of providing feedback to the user. For example, when a user selects an object on a screen, the system should highlight that item so that the user knows that the system has accepted his input and knows how the system

TABLE 8. (*continued*)

has interpreted his input. The latter information is very important when the screen is cluttered and/or the choice is critical for successful subsequent operations [47].

- Important but infrequent events, such as error messages, may need some enhancement or highlighting to be recognized. Place such messages in the user's central field of view [5].
- Labels should be highlighted or otherwise accentuated to facilitate operator scanning and recognition [150].
- The technique used to highlight labels, should be easily distinguished from that used to highlight emergency or critical messages [150].
- Highlighting can be used to provide feedback to the user. For example, when a user is presented with a menu containing a list of mutually exclusive options, and one option on the list is selected, those remaining in the list could be dimmed. This is particularly important when the screen is cluttered.
- A movable cursor with distinctive visual features (shape, blink, etc.) should be used to designate position on a CRT screen. If the position designation involves only selection among displayed alternatives, some form of highlighting might be used instead [5].
- For critical instruction, color could be used to highlight specific ideas or objects [5].
- Important but infrequent events need enhancement by stimuli involving high attention-getting factors. Place such symbols/ messages in the user's central field of view, relative to the display or window [47].
- Refrain from overusing highlighting [53].

Brightness

Brightness (or contrast) enhancement is a useful technique for providing visual distinction between different classes of information. The method simply involves the display of items at different luminance levels. This technique is most often used in the display of alphanumeric information.

Guidelines	**Research Support**	**Source**
1. No more than three levels of brightness coding should be used, with two levels preferred.[a] For example, a data form might combine bright data items with dim labels, to facilitate display scanning.[a]	Y	150, 53, 23, 135, 47, 53, 10, 46, 148, 114
2. brightness coding should be employed only to differentiate between an item of information and adjacent information.	L	16
3. High brightness levels should be used to signify information of primary importance, and lower levels should be used to signify information of secondary interest.	L	57
4. Brightness coding should not be used in conjunction with shape or size coding.[b]	L	46
5. When an operation is to be performed on a single item on a display, the item should be highlighted.	Y	47
6. In a list, the option(s) selected by the user should be highlighted.	Y	47
7. Maximum contrast should be provided between those items highlighted and those not.[c]	Y	47
8. When graphical items are close together on the screen, successive brightening of graphical items and user selection by button activation should be considered.	Y	51

[a] Reference [23] recommends that no more than two brightness levels be used, corresponding to normal and bold appearance; if the application requires additional levels of coding, that should be provided by introducing an additional coding dimension.
References [53], [135], [10], and [148] state that, under optimal conditions, as many as four brightness levels could be used; however, those references also recommend that only two brightness levels be used if accuracy of decoding is a criterion. References [53], [10], [46], and [148] caution that brightness codes can be inefficient because the sorting of different brightness can be distracting and fatiguing; in addition, brighter characters often obscure or mask dimmer characters. Reference [53] further notes that changes in ambient lighting decrease decoding accuracy. Reference [10] qualifies the maximum of four levels by assuming the user is highly trained and frequently uses the code; even with those assumptions, a five prercent error in decoding must be expected.

[b] Shape or size coding influences the apparent brightness of the coded information.

[c] For text, maximum contrast seems to be best achieved by reversing the image (dark on a light background, for example) of the item specified [47].

Comment. All of the above sources tend to be in general agreement concerning the number of coding steps. Two steps are most often recommended. Even though up to four are cited, it is often considered risky to exceed two. If more than two coding levels are needed, Cakir et al. [23] suggest introducing additional coding dimensions. Potential weaknesses in the use of this coding dimension were also noted in the literature:

- Brighter characters often mask dimmer characters, thus reducing legibility.
- Brightness coding, used in conjunction with shape and size coding, may influence the brightness of the coded information. That is probably the foundation for limiting brightness coding to a constant symbolic dimension, such as alphanumeric characters.

- Variations in ambient lighting may hinder the utility of brightness coding.
- Too much visual contrast may cause a dazzling effect to the viewer [45].

Therefore, it is the opinion of the authors that alternative coding dimensions should be considered prior to implementing a brightness coding technique. However, if brightness coding is implemented, due caution should be exercised not to exceed two coding steps.

Blink

Blink coding attracts the attention of the operator to particular items of information by blinking the specific character, label, or field representing the relevant data. This term is also often referred to in the literature as flash rate or intermittency coding. The use and frequency of the blink rate are the most significant parameters associated with defining guidelines for this dimension. Indiscriminate use of blink coding may be an annoyance to the operator. In determining blink rate, one that is too low may not be immediately recognized by the operator. When blink rates are too high, it may be imperceptible by the operator. Therefore, much of the research in this area is devoted to development of optimum blink rates. Guidelines for the use of and determination of effective blink rates are discussed in this section.

	Guidelines	**Research Support**	**Source**
	Blink Rates		
1.	Blink coding should be limited to small fields.[a]	Y	23
2.	The blink rate should lie in the range of 0.1 to 5 Hz, with 2 to 3 Hz preferred.[b]	Y	150, 146, 23, 135, 47,

			10, 39, 120
3.	The minimum "on" time should be 50 ms.[c]	Y	150,47, 120
4.	To avoid interference with reading performance, the blink rate should be such that the user can match the operator's scan rate to the blink rate.	Y	120
5.	If difference in blink rate is used as a coding method, no more than two steps should be used.	Y	150, 53, 23, 135, 47, 10
6.	When two blink rates are used, the fast blink, rate should approximate four per second and the slow rate should be one blink per second.[d]	Y	150, 46
7.	When two blink rates are used, the higher rate should apply to the most critical information.	Y	150, 46
8.	When two blink rates are used, the "on-off" ratio should approximate 50%.	Y	150, 46
	Operator Control		
9.	A means should be provided for suppressing the blink action once the coded data have been located.	Y	146, 23, 53, 114
10.	An "off" condition should never be used to attract attention to a message.	Y	10
	Use of Blink Coding		
11.	Blinking should be reserved for emergency conditions or similar situations requiring immediate operator action.	Y	150,146, 135, 47 53, 46,114
12.	When blink coding is used to mark a data item that must be read, an extra symbol (such as an asterisk) should be added as a blinking marker rather than blinking the item itself.[e]	Y	135

13. Blink coding should be used for target detection tasks, particularly with high-density displays.	Y	114
14. Blink coding should not be used with long-persistence phosphor displays.	Y	120

---

[a] A disadvantage with blink coding is that the blink action may be a source of annoyance to the operator. For this reason, blink coding is more appropriate for small fields and means should be provided for suppressing the blink action once the coded data has been located and is being attended.

[b] Reference [39] states, "Many display designers like to use blinking to provide an attention-getting feature for alarms or other conditions. The human eye responds differently to various blink rates, producing symptoms ranging from nausea to urgency. A blink rate between 0.5 and 2 Hz (once every two seconds to twice per second) is the most attention-getting rate and generates a sense of urgency. Frequencies of 25 to 34 Hz produce symptoms of nausea and can be very annoying. Blink rates below 0.1 Hz (once every 10 s) are basically indistinguishable unless the operator happens to watch the character as it changes."

[c] Reference [120] recommends a minimum on time of 80 ms. References [146] and [135] recommend equal on and off times. Reference [135] further states that an effective code can probably be provided even when the off interval is considerably shorter than the on interval.

[d] Reference [153] state, "Cohen and Dinnerstein [30] trained subjects to identify flash rates varying from one flash every four seconds to 12 flashes every second. Trained subjects were able to use only four rates with reasonably high accuracy. It was found that the maximally efficient frequencies should be equally spaced on a log scale. Because flashing codes are annoying, their application to displays should be limited."
Reference [10] notes that for frequency coding (blink coding), the maximum number of code steps is four with two recommended. The maximum number assumes a high training and use level of the code. Also, five percent error in decoding must be expected. The recommended number assumes operational conditions and a need for high accuracy.

[e] This practice will draw attention to an item without detracting from its legibility [135].

Comment. The following general conclusions can be drawn from the above guidelines:

- Blink rate should be between the range of 0.1 to 5 Hz, with 2 to 3 Hz preferred.
- Blink coding should be reserved for unique attention getting purposes (e.g., emergency alarms) and not overused.
- A maximum of four coding steps are identified, but no more than two are recommended. Code steps beyond this value are risky.
- The operator should be provided with the capability to turn off the blinking after he or she has responded.

In reference to the on/off times for blink coding, a variation among the guidelines was noted. Regardless of the values selected, the following general rule of thumb should apply in all cases: for a blinking display, the cycle should be such that the "on" mode is substantially longer than the "off" mode. This is in contrast to design guidelines that suggest an "off-on" mode of equal intervals (50% on and 50% off) as acceptable [131].

Variations in the use of blink coding have also been examined. Rather than blinking a displayed message, label, or specific item to attract an operator's attention, it seems preferable to present the message in a steady form and indicate its importance by juxtaposing a separate blinking symbol (e.g., flashing arrow) [112].

Image Reversal

Image reversal or reverse video coding is a technique made possible by reversing the polarity of the electron beam so that the beam which would normally be "off" is "on" and vice versa. For example, suppose normal or uncoded data remain displayed as light characters on a dark background (negative polarity). Coded data are presented as dark characters on a light background (positive polarity). Examples of negative and positive image polarity are shown in Figure 14. For many applications, it is assumed that normal data are light characters on a dark background and image reversal coding makes use of the reverse of these parameters (characters and

Xxxx Xxx Xxx

Positive image section
(dark characters on a light background)

Figure 14. (a) Positive image polarity – dark characters on a light background. (b) Negative image polarity – light characters on a dark background.

background). However, there are no conclusive guidelines which state whether normal data presentation should be positive or negative polarity. Arguments in favor of positive polarity (dark characters against a light background) indicate compatibility between the CRT screen and the source document (normally black ink on white background). This premise assumes a high frequency of looking back and forth between screen and source document, such as in a word processing task. The superiority of negative polarity (light characters against a dark background) is supported by the argument that the overall luminance reduction produces less eye strain. Trade-offs between negative and positive polarity warrant further investigation.

The basic point being established in this section is that given a screen of data, whether it is positive or negative, the reversal of desired fields can be an effective coding dimension. However, it will be noted that many of the following guidelines are based on the assumption that negative polarity is normal and regarded as the standard format and that reverse video coding is achieved with positive polarity.

Guidelines		**Research Support**	**Source**
1.	Image reversal (e.g., dark characters on a light background) should be used primarily for highlighting in dense data fields.[a]	L	150, 23, 46
2.	Image reversal can be used to code annunciator information that requires immediate response.[b]	L	135, 46
3.	Maximum contrast should be provided between highlighted and nonhighlighted items.[c]	L	47

[a] Where a large area of the screen is displayed in reverse video, flicker is more likely to be perceived [23].

[b] Ensure that operating sequence codes used in CRT displays are compatible with codes used for annunciator tiles [46].

[c] Provide maximum contrast of a highlighted item with a nonhighlighted item. This seems best done with text by reversing the image (dark on a light background, for example) of the items specified.

Comment. Supporting research for all of these guidelines is extremely scarce. The feasibility of this technique appears to be most appropriate for the highlighting of information in dense fields (e.g., such as words in paragraphs) [150]. The use of image reversal has also been identified in a variety of personal microcomputer software. In that case, image reversal seems to provide more of an aesthetic rather than a highlighting function. In addition to its merits, it should be cautioned that this technique may reduce legibility. Cakir et al. [23] also pointed out that flicker is more likely to be perceived when large areas of the screen are displayed in reverse video.

Auditory

The perception of auditory signals, based on factors such as intensity, frequency, duration, and direction, is especially effective in a variety of applications. There are roughly three types of human functions involved in the reception of auditory signals: (a) detection (determining if a given signal is or is not present, such as a warning signal); (b) relative discrimination (differentiating between two or more signals when presented close together); and (c) absolute identification (identifying a particular signal of some class when only one signal is presented). Relative discrimination and absolute identification can be made on the basis of any of several stimulus dimensions, such as intensity, frequency, duration, and direction (the difference in intensity of signals transmitted to the two ears) [91].

The auditory coding dimension is unique in comparison to other coding techniques. All other enhancement coding dimensions require a stimulus input to the operator through the visual system. In contrast, this particular dimension relies on the auditory senses for detection and interpretation of the stimulus. The use of auditory presentation of information to the operator can often be preferable to visual displays. Some circumstances when auditory coding can be better than visual coding are given below:

- When the origin of the signal is itself a sound,
- When the message is simple and short,
- When the message will not be referred to later,
- When the message deals with events in time,
- When sending warnings or when the message calls for immediate action,
- When presenting continuously changing information of some type, such as aircraft, radio range, or flight-path information,
- When the visual system is overburdened,

- When speech channels are fully employed (in which case auditory signals such as tones should be clearly detectable from the speech),
- When illumination limits use of vision, and
- When the receiver moves from one place to another [91].

McCormick and Sanders [91] caution that these guidelines should be tempered with judgment rather than rigidly followed.

Certain characteristics of auditory coding are sometimes misunderstood. The use of auditory coding techniques is not restricted to alarms and warning devices. Even though auditory signals satisfy many of the guidelines for alarms, their utility should not necessarily be limited to emergency response conditions. This section addresses auditory coding guidelines in a more generic sense. A detailed discussion concerning the use of alarm/warning functions is reported in a separate section. Obviously, this latter section includes auditory coding in addition to a multitude of visual coding parameters which comprise the total alarm/warning system.

Relevant auditory coding guidelines, with particular reference to their suitability in process control applications, are presented below.

Guidelines	**Research Support**	**Source**
Use of Audio Coding		
1. Audio displays should be provided when		
a. The information to be processed is short, simple, and transitory, requiring immediate or time-based response.	Y	146
b. The common mode of visual display is restricted by overburdening; ambient light variability or limitation; operator mobility; degradation of vision by vibration, high g-forces, hypoxia, or	Y	146

other environmental considerations; or
anticipated operator inattention.

c. The criticality of transmission response makes supplementary or redundant transmission desirable.	Y	146
d. It is desirable to warn, alert, or cue the operator to subsequent additional response.	Y	146
e. Custom or usage has created anticipation of an audio display.	Y	146
f. Voice communication is necessary or desirable.	Y	146

Audio System Reliability and Testing

2.	The design of audio display devices and circuits should preclude false alarms.	Y	146
3.	The audio display device and circuit should be designed to preclude warning signal failure in the event of system or equipment failure and vice versa.	Y	146
4.	Audio displays should be equipped with circuit test devices or other means of operability testing.	Y	146

Control Room Environment

5.	Audio displays should be compatible with the ambient conditions in which they are used.	Y	160

Criteria for Audio Signal Selection

6.	If a signal type is commonly associated with a certain type of activity, it should not be used for other purposes when the situation is such that the more common convention is in use.	Y	160
7.	Once a particular auditory signal code is established for a given operating situation, the same signal should not be designated for some other display.	Y	160, 90

8.	The frequency of an audio signal should be within the range of 200 to 5000 Hz, and preferably between 500 and 3000 Hz.[a]	Y	90
9.	When small changes in signal intensity must be detected, the signal frequency should be from 1000 to 4000 Hz.	Y	67
10.	When an audio signal must travel over 1000 ft, its frequency should be less than 1000 Hz.	Y	90
11.	When an audio signal must bend around major obstacles or pass through partitions, its frequency should be less than 500 Hz.	Y	90
12.	When the noise environment is unknown or suspected of being difficult to penetrate, the audio signal should have a shifting frequency that passes through the entire noise spectrum and/or be combined with a visual signal.	Y	160
13.	If a signal must occur in an area in which only certain personnel should be privy to its purpose and others are not to be unduly annoyed, a simple bell tone should be used that is recognizable among ambient speech sounds without being loud.	Y	160
14.	When the signal must indicate which operator (of a group of operators) is to respond, a simple repetition code should be used.	Y	160
15.	Audio signals should not startle listeners, add to overall noise levels, or interfere with local speech activity.[b]	Y	160
16.	Auditory signals should be easily discernible from any ongoing audio input (be it either meaningful input or noise).[c]	Y	90
17.	The intensity of audio signals should be at least 60 dB above the absolute threshold.	Y	17, 90
18.	When complex information is to be presented, two-stage signals should be used.[d]	Y	90

19.	If a person is to listen concurrently to two or more channels, the frequency of the channels should be different.	Y	90
20.	audio signals should provide only that information which is necessary for the user.	Y	90
21.	Where feasible, interrupted or variable signals should be used rather than steady-state signals.[e]	Y	90
22.	Audio signals should be tested prior to using them.[f]	Y	90
23.	Auditory signal frequencies should differ from those that dominate any background noise.	Y	90
	Audio System Modification		
24.	When equipment is modified or added to an existing system, any new audio signals should be compatible with existing audio signals.[g]	Y	90
25.	When an audio signal is installed to replace another type of signal, a changeover period should be allowed during which both the new and the old signal are in effect.	Y	90
	Dedicated System for Warnings		
26.	Where feasible, a separate communication system (such as loudspeakers, horns, or devices not used for other purposes) should be used for warnings.	Y	135, 90

---

[a] The signal frequency of auditory displays should be compatible with the midrange of the ear's response curve for both pitch and loudness, i.e., avoid the use of signals at the extreme ends of the sensitivity curves, where response reliability is more easily masked [160].

[b] Reference [117] states, "Most auditory warning signals are too loud. Most warning signals are designed on the principle that if you want to capture a person's attention, you must hit him over the head with a crowbar."
"Most auditory warning signals are on too long. There are three important consequences: (a) no more than one warning signal can be heard at any one

time; (b) communications are often blocked out in the presence of a continuous on signal; and (c) when manual override is available, operators often disable warning signals."

[c] Reference [117] states, "There are often too many different warning signals. This leads to identification errors, especially in times of stress and emergency."

[d] These stages consist of (a) an attention-demanding signal, to attract attention and identify a general category of information, and (b) a designation signal, to follow the attention-demanding signal and designate the precise information within the general class indicated above.

[e] Use a modulated signal (one to eight beeps/s or warbling sounds varying from one to three time/s), since it is different enough from normal sounds to demand attention.

[f] Such tests should be made with a representative sample of the potential user population, to be sure the signals can be detected by them.

[g] Any newly installed signals should not be contradictory in meaning to any somewhat similar signals used in existing or earlier systems.

Comment. The above guidelines have identified auditory coding criteria for the following concerns: (a) when auditory coding should be implemented (i.e., its merits and limitations) and (b) a general methodology for designing auditory displays. These criteria should provide sufficient background for assessing the suitability of auditory codes in process display environments.

In addition to the guidelines and criteria, Huchingson [67] noted that it is important to actually field-test the auditory codes for detectability and ensure that operators are indeed trained as to their meanings.

Voice

Auditory coding techniques are presently being developed which are not restricted entirely to tonal signals. The technology of voice output provides an emerging design alternative. Computers and associated electronic equipment can produce signals sounding remarkably like words produced by a human voice. In some situations, it may be preferable to have a voice output rather than a CRT display or printout. This dimension will encompass those features of voice coding as a substitute for traditional tonal signals, i.e., preprogrammed signals which respond as a function of plant conditions.

Even though this application is relatively recent in development, there appears to be some potential for its implementation in process control systems. Perhaps the greatest advantage resides in the minimal requirements for learning these codes. On the other hand, immediate problems remain to be resolved in adapting a voice that can be readily accepted by the user.

A review of the guidelines in the area of voice coding and its trade-offs with tonal signals is presented below.

Guidelines	**Research Support**	**Source**
Uses of Voice Communication		
1. A tonal signal should be used when		
a. Immediate action is required on the part of the listeners (i.e., when vocal explanations or directions are not necessary for the listeners to know what the signal means and what they should do).	L	160
b. A specific point in time (that has no absolute value) is to be indicated (e.g., when the sound of a gong tells the listeners that something has happened or is about to happen, that they should be prepared for a message, etc.).	L	160
c. A spoken message would compromise the security of a situation (i.e., when a coded tonal signal would be unrecognizable to persons not privy to the code)	L	160
d. Noise conditions are unfavorable for receiving spoken messages.	L	160

e. Speech channels are overloaded.	L	160
f. A spoken message could annoy listeners for whom it is not intended or when the spoken message could mask other messages.	L	160
g. The intended listeners are familiar with the tonal signal implication or the tonal signal code.	L	160
h. It is desired to use the simplest audio signal.	L	160

2. A spoken message should be used when

a. More message flexibility is needed than a tonal signal can convey.	L	160
b. It is necessary to identify the source of the information.	L	160
c. Listeners have not had training in a special tonal signal code.	L	160
d. There is a need for rapid two-way exchanges of information.	L	160
e. The intended information deals with a future time and when preparation is required (e.g., preparatory to initiating some operation during a countdown).	L	160
f. Use of a tonal signal countdown could result in a miscount.	L	160
g. Operational stress surrounding the intended listeners could cause them to forget the meaning of a tonal signal code.	L	160
h. The message is simple, short, and will not be referred to later.	L	5
i. The message calls for immediate action, visual is already overburdened, or the job requires the user to move about continually.	L	5

Criteria for Voice Signal Selection

3.	For auditory displays with voice output, different voices should be considered for use in distinguishing different categories of data.[a]	L	135
4.	Verbal warning signals should consist of		
	a. An initial alerting signal (nonspeech) to attract attention and to designate the general problem.	L	146
	b. A brief, standardized speech signal (verbal message) which identifies the specific condition and suggests appropriate action.	L	146
5.	Verbal alarms for critical functions should be at least 20 dB above the speech interference level at the operating position of the intended receiver.	L	146
6.	The voice used in recording verbal warning signals should be distinctive and mature.	L	146
7.	Verbal signals should be presented in a formal, impersonal manner.	L	146
8.	Verbal warning signals should be conditioned only when necessary to increase or preserve intelligibility.[b]	L	146
9.	In selecting words to be used in audio warning signals, priority should be given to intelligibility, aptness, and conciseness, in that order.	L	146
10.	Computer speech outputs should be repeatable at user request.	L	5
11.	After each computer speech output, the computer should provide the user the choice of responding with wait, go ahead, or repeat.	L	5
12.	The user should be provided with a means of easily returning to the step in the program	L	5

sequence immediately prior to the computer speech output.

---

[a] At least two voices, male and female, could be readily distinguished, and perhaps more depending upon fidelity of auditory output and listening conditions [135].

[b] Conditioning of signals may be used to increase the strength of cognitive sounds relative to vowel strength. Where a signal must be relatively intense because of high ambient noise, peak-clipping may be used to protect the listener against auditory overload [146].

Comment. The utility of voice coding as an alternative to tonal coding is slowly evolving into a variety of uses. However, its uses should not be spontaneously implemented without addressing its advantages and disadvantages in systematic trade-off analysis.

Audio-Visual

As the term implies, audio-visual warning and signal devices are capable of presenting emergency stimulus information to both the auditory and/or the visual systems. This type of information generally takes on the form of a dangerous condition or a signal that something is or is not operating. Warning devices, as a rule, simply represent two-value information (e.g., on/off, go/no-go). This concept can also be expanded to include cautionary or intermediate information, but the basic function is intended to alert the operator that an emergency event will, or is about to, take place. In the literature, with minor exception, some confusion arises between the terms "warning" and "alarm." The military, as pointed out by Danchak [37], reserves alarms for auditory systems and warning for nonauditory. In industrial applications, all components designed to alert or warn the operator of an emergency action are functionally categorized within the annunciator warning system. However, since no standard definition exists in industry, clarity should not be compromised if these terms are considered to be synonymous in this section.

Before guidelines can be examined, it may be beneficial to further define the function of a good warning device. This can be divided into three basic requirements. It should (a) break through and get the attention of a busy or bored operator, (b) tell him/her what is wrong or what action to take, and (c) allow his or her continued attention to other important duties if this is necessary [148].

The following set of guidelines was identified to ensure that these three requirements can be met in a process control environment. These guidelines, it should be noted, are intended only to supplement the various enhancement coding techniques previously examined. The emphasis will be toward determining top-level human factors guidelines as they apply to the total integrated alarm/warning system. As a result, supporting variables such as blink coding, auditory coding, and color coding are certainly relevant; knowledge of them is essential for assessment of these systems.

The various factors and criteria for design and assessment of alarm/warning systems are extremely broad. Thus, for the sake of being succinct, many of the issues concerned with alarm prioritization, design trade-offs, and overall alarm philosophies have been either deleted or greatly summarized. See References [46], [150], [37], and [6] if further detail concerning design and implementation of alarm/warning systems is desired.

As an aside, it should also be emphasized that a complete review of the alarm systems for the VDUs must be complementary to, and integrated with, the complete control room.

Many of the available guidelines unearthed in this area by the authors have been developed for design and layout of hardwired panels. Standards and guidelines for CRT-type warning/alarm systems are somewhat scarce. Given a VDU in a control room, another limitation of this material is that no criteria exist for functional allocation of alarms between the CRT displays and hardwired annunciator panels. In spite of these limitations, a reasonable proportion of these data is certainly applicable to VDUs and will be presented in this section as deemed appropriate. However, due to the unique flexibility and capabilities of CRT based systems, the direct transfer of these older guidelines should be placed under scrutiny and caution before consideration for use in assessment.

	Guidelines	Research Support	Source
	Supporting Information		
1.	Appropriate trend and status displays for minor upsets should be available to the user.[a]	L	39
2.	Status displays for minor upsets should include a detailed alarm list that identifies alarms by name and number.[b]	L	39
	Alarm Handling Functions		
3.	Errors with a structured response pattern should be handled within the computer and should not trigger alarms.	Y	42, 93
4.	Only those errors that require rapid operator decision-making and intervention should activate the redundant alarms.	Y	42
5.	Early signs of a system going out of specification should be identified by low-level alarms since response time is not as critical.	Y	42
6.	In some operations, it may be desirable to preprogram a hierarchy of alarms that could be altered if changes were made in manufacturing specifications.	Y	42
7.	It should be possible to program the computer to display specific information each time one of the major system failures is detected and triggers an alarm.[c]	Y	42
8.	Information displayed to the operator should include the chronological order of failure.	Y	42

9.	Information that will interfere with decision-making, such as alarms from systems that are secondarily affected by the initial problem, should not be presented to the operator unless requested.	Y	42
10.	Numbers of alarms may be reduced by functional grouping.	Y	56
	<u>Feedback</u>		
11.	Once the operator has responded to an alarm with a controlling action, some feedback should be indicated on the VDU to acknowledge that action has been taken.	Y	42, 52, 116
	<u>Audio Signal Integration with Visual Displays</u>		
12.	When used in conjunction with visual displays, audio warning devices should be supplementary or supportive. The audio signal should be used to alert and direct operator attention to the appropriate visual display.	Y	146
	<u>Criteria for Audio/Visual Signal Selection</u>		
13.	The frequency range of audio warning signals should be between 200 and 5000 Hz and, if possible, between 500 and 3000 Hz.	Y	146
14.	If the user will ever be a considerable distance from the equipment in the performance of other tasks, the signal should be loud but of low frequency (less than 1000 Hz).	Y	5, 146
15.	If the user goes into another room or behind partitions, the signal should be of low frequency (below 500 Hz).	Y	5, 146
16.	If there is substantial background noise, the signal should be of a readily distinguishable frequency.	Y	5, 146
17.	If an auditory signal must attract attention and the above guidelines are inadequate, the signal should be modulated.	Y	5

18.	The frequency of a warning tone should be different from that of the electric power employed in the system, to preclude the possibility that a minor equipment failure may generate a spurious signal.	Y	146
19.	The intensity, duration, and source location of audio alarms and signals should be compatible with the acoustical environment of the intended receiver as well as the requirements of other personnel in the signal areas.	Y	146
20.	Audio warning signals should not be of such intensity as to cause discomfort or ringing in the ears as an aftereffect.	Y	146
	<u>Operator Control</u>		
21.	Controls for operator response to the annunciator system should include silence, acknowledge, reset, and test controls.	Y	150
22.	The alarm should cease <u>only</u> after the user responds appropriately.	Y	5
23.	It should be possible to silence an auditory alert signal from any set of annunciator response controls in the primary operating area.	Y	150
24.	The acknowledgment control should terminate the flashing of a visual tile and have it continue at steady illumination until the alarm is cleared.	Y	150
25.	Acknowledgment should be possible only at the work station where the alarm originated.	Y	150
26.	The reset control should silence any audible signal indicating clearance and should extinguish tile illumination.	Y	150
27.	The reset control should be effective only at the work station for the annunciator panel where the alarm initiated.	Y	150

28. The test control should actuate the audible signal and flashing illumination of all tiles in a panel.	Y	150
29. Periodic testing of annunciators should be required and controlled by administrative procedure.	Y	150
Workstation Layout		
30. To facilitate blind reaching, repetitive groups of annunciator controls should have the same arrangement and relative location at different workstations.	Y	150

---

[a] Reference [39] states, "When minor upsets are either noticed or predicted by the operator, he needs to take a closer look at the process area in question. If trend displays are available for the process area, they can be used to show how key variables have been interacting for the past few minutes. A status display is also useful at this point because it will contain detailed information about the process area."

[b] Reference [39] states, "Alarms can be indicated with reverse video or blinking, and each variable should be represented with a value and an analog bar graph."

[c] Reference [42] states, "Because the operator has to respond quickly to an alarm indicating a major failure in the manufacturing system, it is important to use the computer to assist in the problem-solving process. Building information retrieval into the alarm can halve the time taken to remedy the problematic situation, since most of the response time is spent in gathering relevant information."

Comment. A large portion of recent research and inquiry in the area of alarm/warning systems can best be summarized by Danchak [34], "Once again . . . the operator's ability to cope with large amounts of information has been overtaxed. Process control literature of the 70's repeatedly decries the fact that existing alarm systems are inadequate." Danchak, in addition to several other researchers, substantiates this inadequacy by referring to the sequence of events at the Three Mile Island nuclear power plant. Danchak also recognizes the capabilities of CRTs as potential resources to rectify these problems. "Use of the CRT (to display process alarms) has many advantages in this application, but also many serious disadvantages." It is not sufficient to reproduce the annunciators or error-loggers in electronic form. "One needs to return to basics and answer some fundamental questions, the primary of which is the purpose of an alarm display–CRT or otherwise." The need to go back and address these fundamental issues leads to a research effort conducted by Danchak to answer the following question: "What information do nuclear power plant operators need in order to respond to an alarm?" The results of this analysis consisted of a rank ordering of the information requirements from most important to least important [34]. A listing and description of these information requirements are shown in Table 9.

The major conclusions of these findings are as follows:

- The operators want a simple indication of what the alarm is and as much help as possible in determining the seriousness of the alarm.
- The point English language descriptor satisfies the first criterion, while fields such as current point value/state, violated set point limit, Alarm severity indicator, and perhaps priority satisfy the second.
- Items such as time of occurrence, point identification number, and sequence number are of questionable value.
- Date of occurrence (although useful for later engineering analysis) is obviously something that the operator does not need [34].

Danchak's finding, in conjunction with the criteria identified in the guidelines section and those guidelines for the supporting variables, should provide a relatively comprehensive source of information for determining a thorough assessment strategy.

TABLE 9. INFORMATION REQUIREMENTS FOR AN ANNUNCIATOR ALARM SYSTEM IN ORDER OF IMPORTANCE.(Source: Reference [34])

Rank	Parameter	Description	Example(s)
1.	Point English language descriptor	Name of the parameter in the alarm	Pressurizer pressure, hot leg temperature
2.	Current point value/state	The actual value of the parameter that went into alarm	Pressurizer pressure - 2500
3.	Violated set point limit	The limit that has been violated to cause the parameter to go into alarm	Pressurizer pressure high limit = 2450
4.	alarm severity indicator	A simple designator that specifies which alarm limit set point has been violated	Hihi, hi, lo, lolo
5.	Engineering units	Units associated with any displayed values	degrees F, gallons/ minute, psi
6.	Priority	An indicator that reflects the importance of the parameter when alarmed	Priority 1, Priority 2
7.	Major system designator	Name of the major system of which the alarmed parameter is a part	Coolant, LOOP1,
8.	alarm limits	All possible set points which are associated with a given parameter in alarm	trip set points, hi/lo set points, etc.
9.	Time of occurrence	The hours, minutes, and seconds at which the parameter went into alarm	10:14:56

TABLE 9. (*continued*)

Rank	Parameter	Description	Example(s)
10.	Quality tag	computer generated indicator that reflects the confidence level of the instrument measuring the alarm	Out of range, questionable
11.	Reference	An indicator that tells the operator where he can find more information about the alarm	Display page number, panel containing meter or control
12.	Detector number	The P&ID identifier of the detector which measures the current value in alarm	PCDAXL03
13.	Point identification number	Number used to access the alarmed parameter from the computer	NCP103
14.	Sequential number	A number that indicates the position of the alarmed parameter in the entire alarm list	No. 25
15.	Date of occurrence	The month, day, and year that the alarm occurred	2/24/78

Other Techniques

A variety of "other" methods is available which might be useful as enhancement coding techniques. These techniques range from the simple intuitive techniques with widespread applications to more exotic and unique approaches. A list of such methods includes, but is not necessarily limited to: motion, focus, distortion, graphics (boxes, underlining, etc.), and texture. Many of these techniques are feasible and could be easily implemented into process control applications. Yet, the implementation of others is highly questionable. A cursory examination of guidelines for these alternative techniques not previously mentioned is presented below.

Guidelines		Research Support	Source
	Graphical Techniques		
1.	Graphic coding methods should be used to presentstandardized qualitative information to the operator or to draw the operator's attention to a particular portion of the display.[a]	L	150
	Screen Structures		
2.	Extra spacing, horizontal and vertical lines of differing widths, and perhaps color should be used to set off and highlight data.	L	5
3.	Special symbols (e.g., bullets or arrows) should be used to indicate position and to direct attention.[b]	L	5, 135
4.	Other methods of coding which should be considered for graphic displays and computer-generated drawings include, motion, focus, distortion, and line orientation; on the display surface.[c]	L	23
	Borders		
5.	A border should be used to improve the read-ability of a single block of numbers or letters.[d]	L	42, 53
6.	If several labels, or messages are clustered in the same area, distinctive borders should be placed around the critical ones only.[e]	L	42
	Spacing		
7.	When a special symbol is used to mark a word, it should be separated from the beginning of the word by a space.[f]	L	135

Line

8.	Auxiliary methods of line coding should be considered for graphics applications, including variation in line type (solid, dashed, dotted) and width (boldness).[g]	L	135
9.	When a line is added simply to mark or emphasize a displayed item, it should be placed under the designated item.[h]	L	135, 53

Special Applications

10.	Visual dimensions that should be considered for special display coding applications include variation in texture, focus, and motion.[i]	L	135

Use of Colors to Highlight Data

11.	Related data which are distributed about the screen and data to be updated, etc. should be highlighted in white.[j]	L	18

---

[a] Reference [5] recommends emphasizing objects such as labels, data items, titles, or messages by

- Underlining the object,
- Presenting it in a different style or size font (if the object consists of alphanumeric characters),
- Pointing it out with a noticeably large flashing object (such as an arrow),
- Making a shaded box around the item, or
- Putting graphics (such as a rectangle composed of a string of asterisks) around or near the object.

Reference [5] further states that cartoons can be used to highlight points that call for special attention–such as warnings. They should, however, contain the idea to be communicated and not serve merely as decoration where they might detract from the main intent of a set of instructions.

[b] Symbols chosen for such an alerting purpose should not be used for other purposes in the display [135].

[c] In several applications involving the representation of three-dimensional bodies, the use of lines to represent the three axes and their orientation; can provide a useful depth cue [23].

[d] If space is limited and the character size is critical, it is preferable to fill most of the space within the border. If space is not critical, a larger surrounding border contributes to even better readability [42].

[e] Keep the embellishments to a minimum, since each one reduces the effectiveness of display of the others [42].

[f] A symbol immediately adjacent to the beginning of a word will impair legibility [135].

[g] Perhaps three to four line types might be readily distinguished, and two to three line widths [135].

[h] A consistent convention is needed to prevent ambiguity in the coding of vertically arrayed items; underlining is customary and does not detract from word legibility. For words from the Roman alphabet, underlining probably detracts from legibility less than overlining [135].

[i] Texture can be useful for area coding in graphic displays. Only two levels of focus are feasible, clear and blurred, with the risk that blurred items will be illegible. Perhaps 2 to 10 degrees of motion might be distinguished in display applications where it is an appropriate and feasible means of coding [135].

For more detail, the papers by Barmack and Sinaiko [10] and by Foley and Wallace [50] are recommended.

[j] White is used for highlighting related data which are distributed about the screen and for updates, etc., which are worthy of particular attention [18].

Comment. The utility of this variety of techniques is task-specific and dependent upon the particular application. Experimental research is also limited toward their effectiveness and appropriateness for a given situation. Examples of some of these techniques most appropriate for VDU applications are shown in Figure 15. Some of the more unfamiliar techniques such as focus, motion, and stereodepth appear to possess limited use for CRT applications. In summary, the uses of these various coding dimensions are left up to the discretion of the designer on a case-by-case basis. Without detailed guidelines and supportive research, the greatest misuse of such techniques resides in overuse. For example, if underlining were overused, the following deficiencies could result: (a) its original function as an enhancement technique could be greatly diminished and (b) legibility of the elements would be reduced and the subsequent cluttering effect may

9.0 DIRECT THE SECONDARY OPERATOR SHUTDOWN SECONDARY PLANT EQUIPMENT USING APPENDIX D AS TIME PERMITS.

NOTE:
RCS cooldown must commence no later than one hour after the LOCA occurred. If cooldown must commence immediately go directly to Step 13.0 if this procedure. The following Steps 10.0, 11.0 and 12.0 may be completed as time permits under those circumstances.

10. EVALUATE, RESET AND RESTORE SAFETY SYSTEMS AS FOLLOWS:

CAUTION

SIA MUST BE REINITIATED WHENEVER THE RCS IS NOT AT LEAST 20°F SUBCOOLED.

CAUTION

IF RCS ACTIVITIES ARE EXCESSIVELY HIGH, DO NOT RESTORE LETDOWN.

Figure 15. formats that highlight cautions, warnings, and notes using boxes, underlines, and different margins.

actually degrade performance. Unfortunately, no data are available to determine what represents "overuse." Therefore, each screen must be subjectively evaluated on a case-by-case basis to ensure that whatever technique is employed, it is used parsimoniously.

Dynamic Display

Many CRT displays are dynamic in nature. That is, they are continuously driven by various data inputs to update the condition of the displayed plant or process parameters. The determination of the rate of these changes before overload or loss of comprehension is an area in need of further research [7]. The rate of change is, to some extent, related to the nature of the process. Dynamic, critical processes will require more frequent updating.

In this subsection, display motion and digital counters are examined as methods of dynamic display presentation.

Display Motion

Display Motion refers to the degree of movement or animation present in a given display, e.g., water moving through a pipe that is dynamically and graphically presented, control rods moving up or down, or a valve changing position from open to closed. All of these examples indicate some dynamic movement or change that could be presented graphically to operators.

One of the important questions associated with animated display motion is how fast or slow should the apparent motion of the graphic representation be projected to the operator. The display should clearly be of sufficient magnitude to be easily detected and recognized by operators. If the rate of pixal excitation is too great, a progressive incremental movement perception will not be readily apparent to the operator, but instead a "zip effect" will be produced. If one were to prescribe an animated fluid flow rate linearly in terms of pixels per second, such a definition would be unsatisfactory because pixel size and dimension are not standardized within the computer industry. A better definition, one which would enable the easy establishment of guidelines and the applicability to virtually all types of CRTs, would be to state rate in terms of millimeters per second.

Unfortunately, there are no specific guidelines available for determining these rates. However, some basic building blocks toward developing such guidelines have been examined. These developments are the product of both observational analysis and formal experimentation. The results of these studies, in addition to relevant top-level design criteria, will be summarized in the guidelines sections below. The reader is referred to References [80] and [128] for further information.

	Guidelines	**Research Support**	**Source**
1.	The speed of a graphic showing fluid flow in a pipe should be greater than 7.28 mm/s (0.29 in./s) but less than 295 mm/s (11.8 in./s).[a]	L	7
2.	Changing values which the operator uses to identify rate of change or to read gross values should not be updated faster than 5 s nor slower than 2 s when the display is to be considered as real time.[b]	L	146
3.	A display freeze mode should be provided to allow close scrutiny of any selected frame.[c]	L	146
4.	Display formats should be designed to optimize information transfer to the operator by means of information coding, grouping, and appropriate information density.	L	146
5.	Rate of motion should not exceed 60 deg/s of visual angle change with 20 deg/s preferred.[d]	Y	53, 125

[a] In informal unpublished studies conducted at the INEL facility in 1980 by Banks, some rough guidelines were generated that were not rigorously tested. To display fluids moving through a piping vessel, a value of 7.28 mm/s (0.29 in./s) continuous flow was found to be the approximate slowest rate of apparent motion to be comfortably detected by several LOFT operators (with

operators (with operator's viewing distance ~24 in. from screen). When values exceeded 295 mm/s (11.8 in./s), operators reported that the fluids moved too fast [7].

[b] Graphic displays requiring operator visual integration of rapidly changing patterns are an exception and should be updated at the maximum refresh rate of the display device consistent with the operator's inormation handling rates [146].

[c] An option should be provided to allow resumption at either the point of stoppage or at the current real-time point. The operator shall be warned if an important event occurs while the display is frozen. An appropriate feedback label should be provided to remind the operator when the display is in the freeze mode [146].

[d] Dynamic visual acuity (DVA) is generally defined in terms of the smallest detail that can be detected when the target is moving. Angular movement of the target decreases the threshold of visual acuity; [153], [25]; DVA also varies with age [125].
Reference [125] states, "Loss of DVA increases rapidly as the rate of motion exceeds 60 deg/s. Such thresholds could be considerably higher with a shorter viewing time or target travel distance.[20], [95]. At rates of 20 deg/s, our ability to resolve detail is almost as good as with motionless targets."

Comment. With the exception of some general findings cited in Van Cott and Kinkade, [153] formal laboratory experimentation directly applicable to minimal safety-related standards or guidelines are limited. The findings of Banks [7] represent the only known relevant guidelines for this variable in process control systems. It should, however, be cautioned that these data are only observational and not supported with formal research. There is an obvious need to perform research in this area to anchor any suggested guideline to laboratory validation and minimize reliance on intuition and casual observation.

The general design guidelines stated in MIL-STD-1472C [146] are also vague but should provide the assessor with some usable benchmark criteria. A display freeze mode as recommended in that standard seems to be a beneficial resource to the operator.

Digital Counters

Another aspect of dynamic characteristics involves the display of quantitative, and often real time, information through a counter-type format. Though this mode of information presentation is often more accurate than scales, it is also subject to gross reading errors. A possible contributing factor to these reading inaccuracies is the counter's rate of change. An overview of the general guidelines associated with establishing these values is presented below.

Guidelines	Research Support	Source
1. Numerals should not follow each other faster than 2/s when the operator is expected to read the numerals consecutively.	L	150
2. Changing values which the operator must reliably read should not be updated faster than 1/s, with a 2-s minimum time preferred.	L	150, 146

Comment. In contrast to MIL-STD-1472C, [146], a more conservative update rate is set forth in NUREG-0700 [150]. A 2-s minimum value is recommended, due to the reduced safety margin established with a longer reading time. In addition, for counters, the following features should also be incorporated into the overall design criteria: (a) multi counters should be oriented to read horizontally from left to right (b) simple character fonts should be used; and (c) where more than four digits are required to display numerical values, the digits should be separated into groups and the groups separated by commas, a decimal point, or additional space.

Information Formats

The flexibility of a CRT as an operator interface provides the display designer with a multitude of formats to select from. Historically, feasibility and development costs precluded all but the most conventional display types. This section investigates both traditional displays and a range of effective graphical techniques currently being proposed for use in VDU technology.

The traditional displays and techniques discussed in this section are

- Analog
- Digital
- Binary indicator
- Bar/column chart
- Band charts
- Linear profile
- Circular profile
- Single value line chart
- Trend plot
- Mimic display

The particular advantages, uses, and restrictions of each presentation technique or method are discussed in this subcategory.

The massive number of these display patterns can be overwhelming to the designer who is tasked with the mission of selecting the most optimal design. Danchak [35] surveyed the literature concerning the available techniques for display of multivariate data and uncovered over 60 methods of graphical representation. Fortunately, the methods for selecting a suitable format are not entirely subjective. A man/machine systems engineering approach provides the basic elements for determining the fundamental requirements for the designer to base his or her decision. Even though a complete systems analysis is the most effective strategy, the concepts can still be entertained on a more informal scale. Before selecting a display, the designer must possess or have access to the following information [45]:

- For each task
  - What are the plant parameters that the task depends on?
  - What are the objectives to be accomplished with the use of the display system and how will the operator benefit from it?
  - Can the set of plant parameters for this task be used in an absolute or relative manner?
- For each plant parameter to be displayed
  - Is quantitative or qualitative information required?
  - What should the range of each parameter be?
  - What should the required level of accuracy for each parameter be?

In short, before selecting a particular display type, the designer should consider both the type of data and the intended use of the display [46]. Once these basic information requirements are fulfilled, the designer can proceed with selection of the specific information or display pattern.

As mentioned previously, the number of available display patterns is enormous. However, a great deal of preliminary analysis has already been conducted (based on Danchak's findings) to determine a useful subset of candidate displays [151], [45]. This candidate group was founded on the determination of the most feasible (and, in some cases, not feasible) types proposed for implementation in process control applications. One list of display options was generated by Electric Power Research Institute (EPRI) [45]. A second list identified by the author is located in NUREG-0835

[151]. The results of those lists were summarized and collated to produce the display categories presented herein.

A set of general guidelines and design principles applicable to all information formats is shown in Table 10. This top-level table supplements the guidelines and criteria presented for each specific format. These guidelines should be reviewed prior to investigating the specific variables.

TABLE 10. GENERAL GUIDELINES AND PRINCIPLES FOR INFORMATION FORMATS

Design Requirements

- Variety. A limited variety of graphic display techniques should be used, so that personnel may become sufficiently familiar with each to extract information rapidly from the display.
- Content. The information displayed to an operator shall be limited to that which is necessary to perform specific actions or to make decisions.
- Precision. Information shall be displayed only within the limits and precision required for specific operator actions or decisions.
- Format. Information shall be presented to the operator in a directly usable form.
- Redundancy. Redundancy in the display of information to a single operator shall be avoided unless it is required to achieve specified reliability.
- Combining Operator/Maintainer Information. Operator and maintainer information shall not be combined in a single display unless the information content and format are well suited to, and time compatible, for both users.
- Display Failure Clarity. Failure of a display or its circuit shall be immediately apparent to the operator.
- Demand Information Versus Status Information. Demand information shows that equipment has been commanded (by control settings or otherwise) to a particular state or level. It

Table 10 (*continued*)

shows only what is demanded–not what is actually being realized. Status information shows the state or level actually in effect.

- To prevent operator confusion, it is essential that displays be identified as to whether they reflect demand or actual status.
- Video display of actual system/equipment status should be displayed for all important parameters.

<u>Selection Criteria</u>

- Use the simplest display concept commensurate with the information transfer needs of the operator or observer. The more complex the display, the more time it takes to read and interpret the information provided by the display and the more apt the observer or operator is to misinterpret the information or fail to use it correctly.

- Use the least precise display format that is commensurate with the readout accuracy actually required and/or the true accuracy that can be generated by the display-generating equipment.

  Requiring operators to be more precise than necessary only increases their response time, adds to their fatigue or mental stress, and ultimately causes them to make unnecessary errors.

- Use the most natural or expected display format commensurate with the type of information or interpretive response requirements. Unfamiliar formats require additional time to become accustomed to, and they encourage errors in reading and interpretation as a result of unfamiliarity and interference with habit patterns. When new and unusual formats seem to be needed, consider experimental tests to determine whether such formats are compatible with basic operator capabilities and limitations and/or whether the new format does in fact result in the required performance level.

Table 10 (*continued*)

- Optimize the following display features:
  - Visibility: Viewing distance in relation to size, viewing angle, absence of parallax and visual occlusion, visual contrast, minimal interference from glare, and adequate illumination.
  - Conspicuousness: Ability to attract attention and distinguishability from background interference and distraction.
  - Legibility: Pattern discrimination, color and brightness contrast, size, shape, distortion, and illusory aspects.
  - Interpretability: Meaningfulness to the intended observer within the viewing environment; requirements for interpretation, extrapolation, special learning, and training; and general reliability in terms of retention of meaning.
- Graphics and Labels. Illustrations, line drawings, and animation should be used to supplement the explanations in the text. Graphics are especially useful for spatial visualization problems or where the problem to be solved has multiple interacting dimensions. Graphical dialogues are intrinsically motivating, at least for the novice user.
- The axes of graphs should always be labeled.
- Labels for graphs should describe what is being displayed, not the name of the display.
- The axes of graphs should be subdivided approximately with divisions of 1, 2, 5, or 10, not with 3, 7, or other numbers obtained arbitrarily through division.
- The number of graduation marks between numbered scale points should be greater than nine.
- Scales should be numbered starting at zero.
- Increases in magnitude on scales should be clockwise, left to right, or bottom to top.

TABLE 10 (*continued*)

- Maximum contrast should be maintained between scale markings and the value displayed.
- The size of the letters used to label scales should be independent of the scales on the displays on which they appear. If a user contracts a graphic display, labels must not be contracted to the point where they cannot be read.
- If trend lines are to be compared, multiple lines should be used on a single graph.
- Symbols should be designed with consideration of the graphic conventions to which the user may be accustomed, while at the same time being as economical as possible in the use of screen space and image complexity.
- Unnecessary ornamentation, unwanted graphic patterns and illusions, and flaws in alignment should be avoided in graphic displays.
- In graphic displays, the center of rotation should be the center of the object.

Source: The material in this table was developed from information in References [151], [46], [146], [150], [160], [47], [120], [87], [10], [126], [106], [51].

Analog

This type of information format is best characterized by a meter configuration (the most common being a moving pointer/fixed circular scale) but can also take the form of curved (arc), horizontal (straight), and vertical (straight) arrangement of either fixed pointer/moving scale or vice versa. Regardless of arrangement, values are indicated to the observer by a pointer position to scale relationship. Examples of moving-pointer and fixed-scale indicators are shown in Figure 16.

The traditional uses of this format originate from hardwired instrumentation, but software-generated meters which rely on these previous conventions and guidelines are certainly plausible for CRT applications.

General guidelines related to design and assessment of this information format are presented below. For the interested reader who desires further information concerning these formats beyond the guidelines discussed herein, NUREG-0700, [150], MIL-STD-1472C, [146], and Van Cott and Kinkade [153] are recommended.

Guidelines	Research Support	Source
1. Analog displays should not be used when quick, accurate readings are a criterion.	Y	45
2. Numbers should increase clockwise, left to right, or bottom to top, depending on the display design and orientation.	Y	67
3. For one-revolution, circular scales, zero should be at 7 o'clock and the maximum value should be at 5 o'clock, with a 10-degree break in the arc.	Y	67
4. When check-reading positive and negative values, the zero or null position should be at 12 o'clock or 9 o'clock.[a]	Y	67
5. All numbers should be oriented upright.	Y	67
6. Numbers should be outside the graduation (tick) marks unless doing so would constrict the scale.[b]	Y	67
7. The pointer on fixed scales should extend from the right of vertical scales and from the bottom of horizontal scales.	Y	67

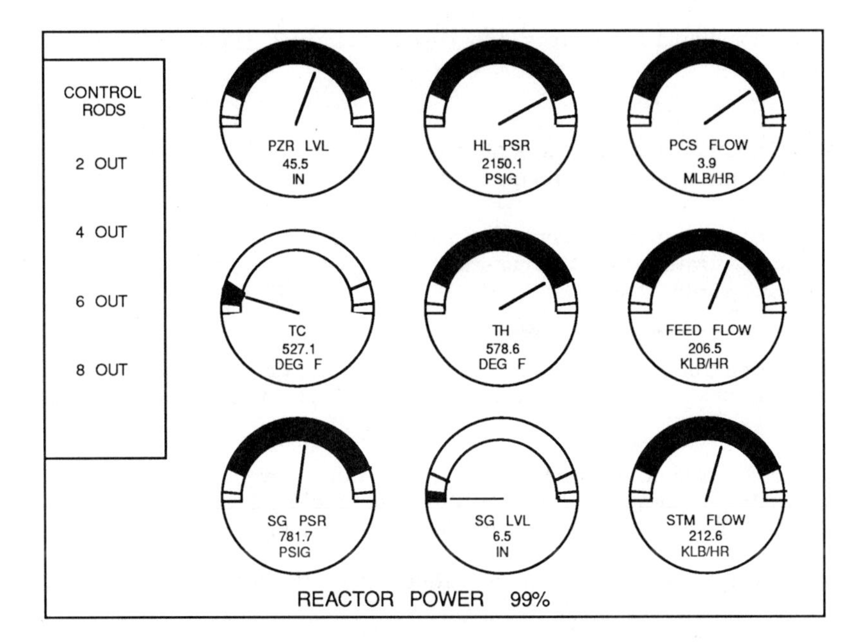

Figure 16. A CRT display using moving pointers and fixed scale indicators.

8.	The pointer on fixed scales should extend to but not obscure the shortest graduation marks.	Y	67
9.	Graduation interval values of fixed scales should be 1, 2, 5, or decimal multiples thereof. Numbering by 1, 10, or 100 is recommended for progressions.	Y	67
10.	Nine should be the maximum number of tick marks between numbers.	Y	67
11.	Tick marks should be separated by at least 0.07 inches for a viewing distance of 28 inches (710 mm) under low illumination (0.03-1.0 fL).	Y	67
12.	Dials should not be cluttered with more marks than necessary for precision.	Y	67

13.	Zones should be colorcoded by edge lines or wedges. Red, yellow, and green should be used.[c]	Y	67
14.	Information should be in a directly usable form (for example, percent, RPM).	Y	67
15.	Fixed-scale, moving-pointer displays should be used rather than moving-scale, fixed-pointer displays.	Y	150

[a] With a matrix of circular displays, deviations from a 9 o'clock null position are easily detected in check reading. Zero is at 12 on multirevolution dials [67].

[b] Although this Guideline is desirable (for preventing pointer head obstruction of numbers), the crowding of markings on small dials has resulted in most numbers being inside the marks [67].

[c] Zones can be used to indicate operating ranges, dangerous levels, etc [67].

Comment. Conventional analog displays are exceptional for specific applications. The general pointer orientation; gives a quick cue to scale-pointer relationship, rate of change, and reference to scale limits [160]. This technique is also suitable for qualitative information, check reading, for setting in numbers, and for tracking [67]. The disadvantage is its requirement of large panel space. It can be argued that this technique does not fully exploit the full graphic capability of computer technology. However, such a display arrangement is readily familiar to operators, thus minimizing training and potential for confusion. [45]

Digital

This information format represents the simplest and most frequently implemented display type for process control. Digital displays (counters) are most often recommended when presentation of precise quantitative data (or exact value) is required. They are best suited to one dimension, one variable, and one value, but they may be clustered together to show multiple dimensions or variables [45]. An example of this information format is shown in Figure 17.

An overview of the relevant guidelines pertaining to digital displays is presented below. As with many of the traditional display devices of this type, the majority of such guidelines is for mechanical counters rather than electronic or CRT displays. However, many of the basic principles from these earlier guidelines for mechanical devices are certainly applicable.

Guidelines	Research Support	Source
1. Each display should have a label to identify its meaning.	Y	45
2. Digital displays should include the appropriate number of significant figures for the required level of accuracy.	Y	45
3. Digital displays should accommodate the full range of the variable (i.e., highest and lowest values).	Y	45

PZR LVL	44.50	In
PCS flow	3.80	Mlb/hr
Feed flow	142.80	klb/hr
STM flow	141.60	klb/hr
SG LVL	7.30	In
T steam	525.90	°F

Reactor power – 75%

Figure 17. Examples of digital/counter type displays.

4.	Digital displays should change slowly enough to be readable.	Y	45
5.	Digital displays should be provided with arrows to indicate the direction of change (if that is likely to be needed).	Y	45
6.	If more than four digits are required, they should be grouped and the groupings separated as appropriate by commas, a decimal point, or additional space.	Y	150
7.	Multidigit counters should be oriented to read horizontally from left to right.	Y	150
8.	Simple character fonts should be used.[a]	Y	150
9.	Horizontal spacing between numerals should be between one-quarter and one-half the numeral width.	Y	150

[a] Styles using variable stroke widths, slanted characters, etc. should be avoided [150].

Comment. Many of the above guidelines are echoed throughout the human engineering design literature and appear relatively standardized across references. One exception was noted in the guidelines between MIL-STD-1472C [146] and NUREG-0700 [150]. NUREG-0700 recommends, "If more than four digits are required, they should be grouped and the groupings separated as appropriate by commas." MIL-STD-1472 states: "Commas shall not be used." The trade-offs are not altogether clear, whether pro or con, for commas. Consideration for this guideline is probably dependent on a task-specific situation.

In the guidelines concerning digital displays, Bailey [5] also noted some disadvantages to this format: (a) determining the rate of change is difficult; (b) reading rapidly changing display values is difficult; and (c) gauging distance to a boundary (control limit, danger zone) is difficult.

These constraints should be examined in conjunction with the design guidelines to ensure that this format is the most appropriate technique for a given application.

Binary Indicator

This technique and the guidelines proposed for it are representative of another information format often used on conventional display panels. It is usually applied to meet those display requirements which require two-valued information (go-no go, start-stop, safe-unsafe, warning-caution, etc.). The most common of these uses is seen in alarm/warning devices. An example of this information format is shown in Figure 18.

A brief overview of this technique is discussed in the guidelines below. Further information with specific reference to this format as an alarm/warning device can be gleaned from the section on enhancement coding.

PRIMARY FLOW	COLD LEG TEMPERATURE	DELTA TEMPERATURE	PRIMARY PRESSURE	PRESSURIZER LEVEL
SECONDARY PRESSURE	SECONDARY FEED FLOW	STEAM CONTROL VLV POSITION	STEAM GEN LEVEL	CONDENSER PRESSURE

Figure 18. Example of alarm/warning device display.

Any assessment of this information format would not be complete without consideration of the assorted coding dimensions associated with this type of format (flash rate, color, intensity, etc.). In addition, as with other alarm/warning techniques, this method should not be overused. Too many indicators will clutter the screen or make it appear busy, thus overloading the operator.

Guidelines	Research Support	Source
1. Binary indicators should be clearly labeled and understood.	Y	45
2. For quantitative measurements, binary indicators should be used only for check-reading purposes.	Y	91
3. Where meaning is not apparent, labeling should be provided close to the status indicator.	Y	150
4. When monochrome is not used, the color of the indicator should be clearly identifiable.	Y	150
5. Symbolic legends should be clear and unambiguous as to their meaning.	Y	150
6. Legend text should be short, concise, and unambiguous.	Y	150
7. Legend nomenclature and abbreviations should be standard and consistent with usage throughout the control room and in the procedures.	Y	150
8. Legends should be worded to tell the status indicated by the display.	Y	6
9. The legends of illuminated indicators should be readily distinguishable from legend push buttons by form, size, or other factors.[a]	Y	150

[a] This criterion has special emphasis to VDUs when touch panels are employed as a control device.

Comments. An assessment of this information format would not be complete wthout consideration of the assorted coding dimensions associated with this type of format (flash rate, color, intensity, etc.). In addition, as with other alarm/warning techniques, this method should not be overused. Too many indicators will "Clutter" the screen or make it appear "Busy," thus overloading the operator.

Bar/Column Charts

The bar/column chart is probably one of the most popular graphical information formats proposed for use in process control. The different arrangements of displays which implement a bar chart approach are almost limitless. For bar charts, simple and deviation are two techniques most often encountered. A simple bar chart contains horizontally oriented rectangles or bars emanating from a single vertical line. In a deviation bar chart, each item has a bar extending either to the right or left of a common vertical base line to indicate deviations from some normal value [35].

Similar configurations exist for column charts, with the difference being vertically oriented rectangles. This seemingly trivial difference leads to subtle differences in how information is perceived. First, because the column heights move vertically, there is a natural compatibility between a rise in the value of a variable and a rise in the column height. Second, because the scale of the variable is on the Y axis, the X axis can be used to denote time. (For a bar chart, time would be on the Y axis, which runs counter to convention.)

In the conventional column chart, values increase from bottom to top [45]. This technique, as described in NUREG-0835 [151], "synthesizes an array of analog meters." Further definition can be best described by the following attribute germane to all bar/column charts: the length of each bar is generally proportional to the magnitude of the measured parameter it represents [151]. Examples of simple bar, simple column, and deviation bar charts are shown in Figures 19, 20, and 21.

The guidelines in this section are devoted to these major types with the known intent that a multitude of variations can be founded on the basic configurations discussed here.

Guidelines	Research Support	Source
1. Each bar on the display should have a unique identification label.[a]	Y	151
2. Bar charts should contain reference(s) to the normal operating condition(s).[b]	Y	151
3. Column charts should be used when the direction of change of the measurement is to be emphasized or when time is represented by one of the axes of the chart.	Y	45

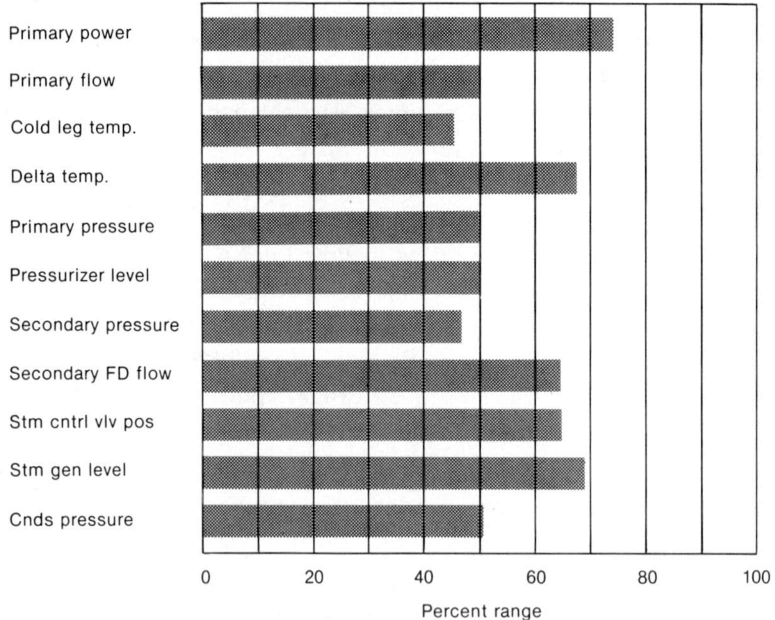

Figure 19. Example of a simple bar chart.

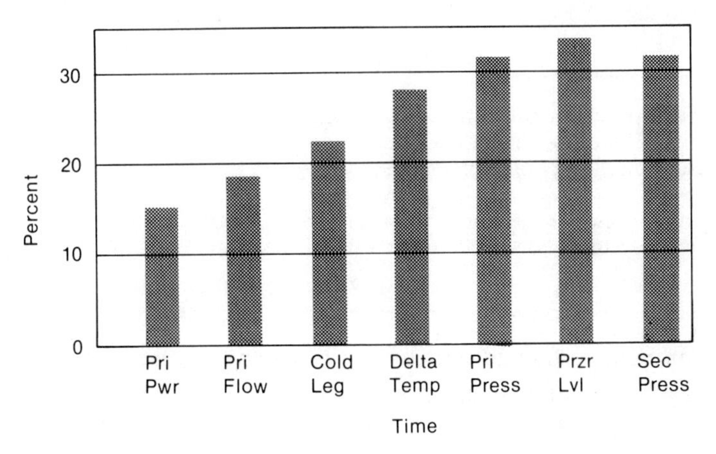

Figure 20. Example of a simple column chart.

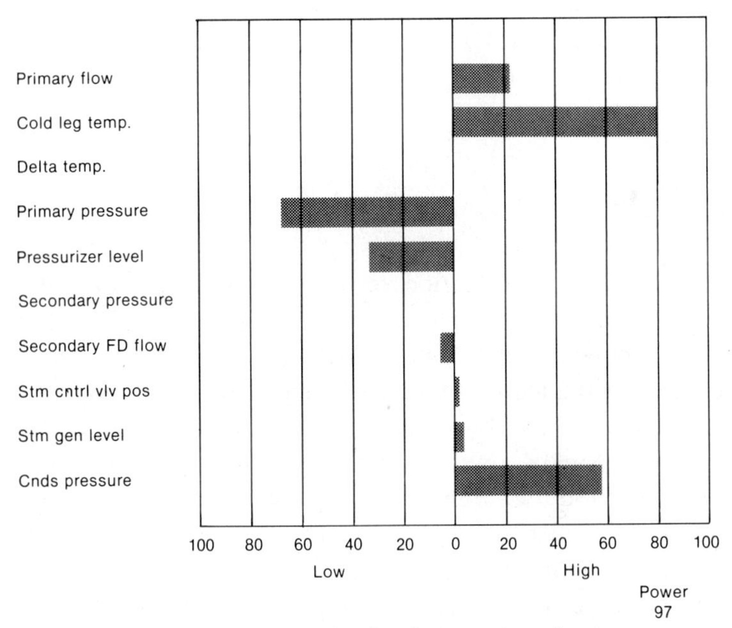

Figure 21. Example of a deviation bar chart.

4.	Stroke type charts are alternatives to conventional full bars.	Y	155

---

[a] The label provides a positive identification of the parameter each bar represents. It would not be acceptable for an operator to have to memorize the position of each parameter on the display [151].

[b] With references showing normal parameter operating values, the operators are more likely to notice deviations from normal conditions [151].

Comment. In addition to the guidelines presented above, the displays should also comply with good human factor engineering standards and practices, i.e., are the displays legible, cluttered, inconsistent, labeled, etc. It is also important to emphasize the need for reference points to indicate normal operating status. In other observations of these guidelines, it was noted that they have failed to discuss the advantages or disadvantages between column versus horizontal presentation. The only known guidelines addressing this issue were located in work performed by EPRI [45], but detailed trade-off guidance between horizontal and column presentation is limited.

Band Charts

This type of information format contains a series of bands depicting the components of a total series. The values of the bands (or strata) are plotted on an X-Y plot. Each of the bands are added to one another so that the topmost boundary represents the sum of all bands. A report prepared by the Electric Power Research Institute states, "Band charts can be used to show, for example, how much each turbine is contributing to total flow. This format is most useful when all elements contribute equally to the total under normal circumstances." [45].

An example of this type of information format is presented in Figure 22.

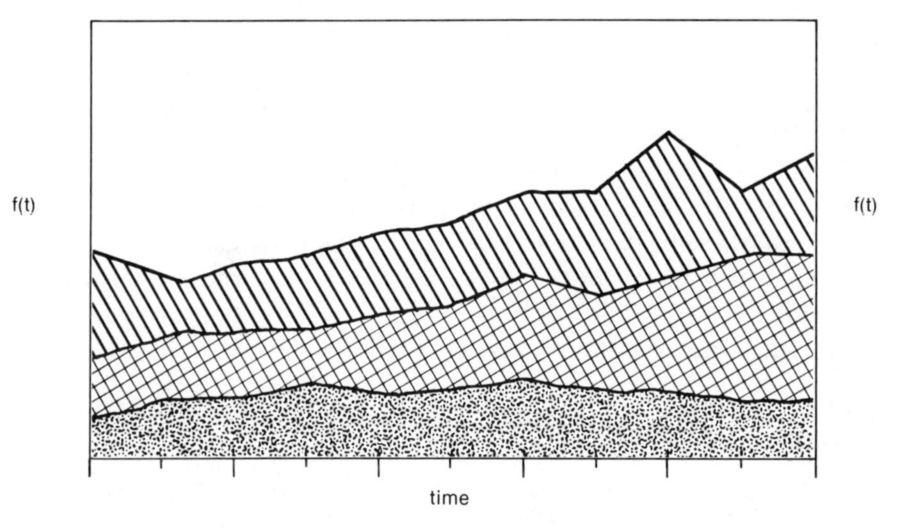

Figure 22. Example of a multiple surface/band chart.

Guidelines	Research Support	Source
1. All items on a band chart should be related to the total.	L	35

Comments. Specific guidelines pertaining to this format are limited. However, this format appears to have useful application provided its implementation does not violate the following constraints identified by Danchak [35]:

- Not to be used when changes in the movement of a series are abrupt.
- Not to be used where accurate reading of a component is of paramount importance.

The layout of the display type should also be in compliance with good human factors engineering standards and guidelines.

Linear Profile

The linear profile format is analogous to a modified version of a column chart and band chart. The wide bars common to the column chart are replaced by thin lines. The vertical height or magnitude of those lines, all emanating from a baseline on the X axis, represent parameter values. The maximum heights of each line are connected by a profile line (analogous to a band chart). This profile line graphically shows the nature of a relationship between variables. For example, an irregular profile would be indicative of abnormal operating conditions.

An example of this type of information format is shown in Figure 23.

Guidelines	**Research Support**	**Source**
1. A horizontal line representing normal operating conditions should be superimposed on the display.[a]	Y	151

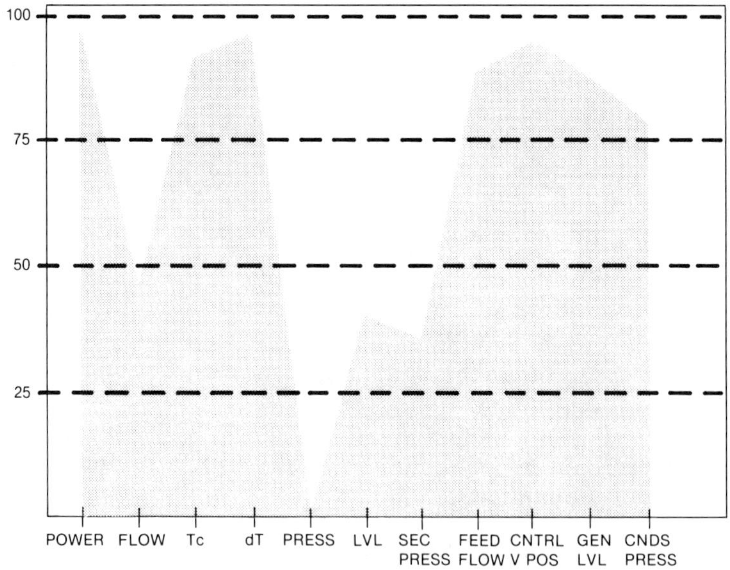

Figure 23. Linear profile display.

2.	The area below the profile line should be shaded to provide a more distinguishable profile.	Y	151
3.	Labels should be provided along the bottom to identify each parameter.	Y	151
4.	Linear profile charts should be used in applications where detection of abnormal events is important.[b]	Y	135

[a] Scaling considerations for the linear profile are the same as for the bar chart. A horizontal line, representing normal operating conditions superimposed on the display, is an acceptable enhancement [151].

[b] The linear aspect of this display is confusing and makes pattern recognition marginal, but it is considered adequate for determining the problem as well as severity [35].

Comment. This format provides the operator with a set of pattern recognition cues which would be highly desirable for detection of abnormal events. Though it is not considered as effective as circular profiles in which the patterns generated are deemed more prominent, Danchak [35] also suggests that the simple bar chart design may be superior to linear profiles. As evidenced from the above guidelines, specific data for this format are lacking and more research is needed.

Circular Profile

This format is also often referred to as a star display or n-fold circular display. In a circular profile, the same principles apply as with the linear profile with one distinct difference: the parameter lines of a linear profile, which emanate from a common baseline, are joined at a common origin. In this manner, the parameter lines (or spokes) now radiate from the origin point (center of the circle) with equal angular spacing relative to each other. As with the linear profile, the distance along each spoke from the center represents the value of the parameter, and the scales are indexed such that the normal value of each of the variables can be connected to form some easily recognizable polygon. For example, under normal operating conditions, the profile should be circular. An irregular profile is indicative of an abnormal operating condition.

An example of this information format is shown in Figure 24.

Guidelines	Research Support	Source
1. The chart should be designed so that it forms recognizable geometric patterns for specific abnormal conditions.[a]	L	45
2. Labels should be provided to identify each radial line.	Y	151
3. The area within the profile should be shaded to enhance the operator's perception of plant status.	L	151

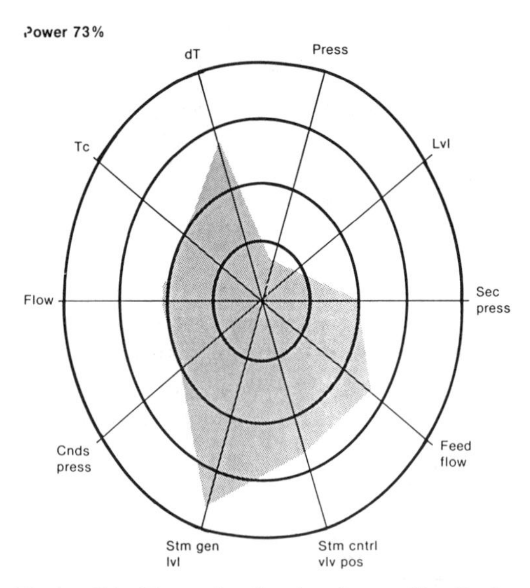

Figure 24. Example of a circular profile display.

[a] Reference [45] reports, "As the user becomes experienced with the circular profile, the asymmetrical polygons that result from off-normal situations should become more familiar. For example, a steam generator tube rupture may result in an hourglass shape or a loss-of-coolant accident might produce a cloverleaf design. Research with this type of picture element has shown promise but may require more training than is necessary for use of the more familiar picture elements."

Comment. This format is considered an extremely effective mode of providing a set of pattern recognition cues to the operator as an aid for detection of abnormal events. It is also useful for showing the relationship between variables that do not have the same units.

As evidenced from the guidelines, specific data for this format are lacking and more research is needed.

Single Value Line Chart

This format is a parameter-versus-parameter-type chart where one dimension is plotted against the other (e.g., pressure versus temperature) on an X-Y axis. This generic pattern is well suited for design of the classical pressure-temperature (P-T) map used in the nuclear industry. The P-T map concept is not a recent innovation. The principles of this technique have historically been implemented as an operator information aid in a hardcopy format. Such P-T graphs are integrated into the plant operating procedures and technical specifications. The recent innovations in VDU technology have made it possible to display dynamic process information in this form. The basic P-T map has the following features:

- Parameter values (e.g., hot leg and cold leg temperature and pressure) are plotted as an identifiable cursor on the map.
- A saturation curve is provided which applies to both primary and secondary water and steam conditions. Above the saturation line is the subcooled water region; below it is the superheated steam region.

- Each of the cursors is surrounded by a box which represents normal operating boundaries. Deviations from those normal operating boundaries are indicated by data points (or trails).

An example of this information format is shown in Figure 25.

Guidelines	**Research Support**	**Source**
1. The target area should be defined.[a]	L	45
2. Old data points should be removed after some fixed period of time.[b]	L	45

[a] Reference [45] reports, This sort of display is best used for detecting deviations from normal if a target area can be defined. By plotting a brief time history, one may be able to predict where the values are headed. Care should be taken to distinguish the current value from past values, especially when the values change slowly. This can be done by placing an X or some other mark at the current value."

[b] Reference [45] states, "To avoid clutter, the old data points should be removed after some fixed period of time. Ideally, as one new point is plotted the oldest point is removed, thereby maintaining a constant number of displayed points. Hardware limitations may force one to plot an entire series, clear the screen, and begin plotting over again, although this is much less desirable."

Comment. When effectively applied, an information-rich display can be created using a P-T map concept. The introduction of the data trails also aids the operator in mitigating an abnormal event. Therefore, the P-T map can be used to not only detect and identify an event but to track and follow to mitigation as well. However, caution should be exercised to avoid the display of too much information. Otherwise, clutter is induced, and operator performance may actually be degraded. Other potential problems may reside in the training requirements needed to effectively utilize this information resource. In other words, operators cannot be expected to do well if training is either cursory or inadequate.

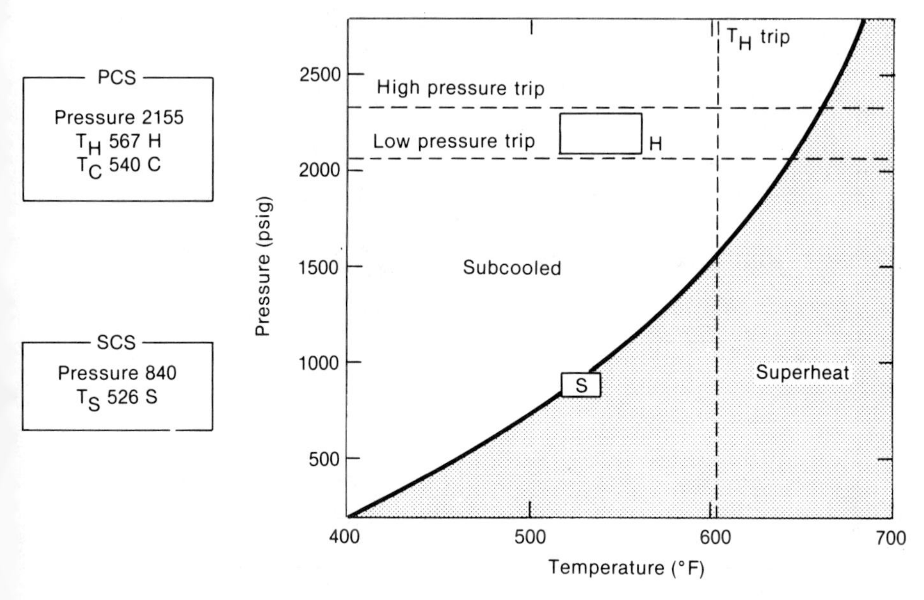

Figure 25. Example of a single value line chart.

Trend Plot

This information format is simply an X-Y axis with the behavior of the X-Y values described by a function curve. Time is generally plotted on the X axis, while the Y axis is reserved for some parameter value. The plot historically depicts how one or more variables and/or one or more dimensions vary over time.

An example of this information format is presented in Figure 26.

A special application of the trend plot technique that also makes use of bar charts and digital displays is shown in Figure 27.

Research has shown that if time is plotted on the X axis, it should increase from left to right (i.e., the most recent data on the right). However, if time is plotted on the Y axis, it should increase from top to bottom, toward the origin, (i.e., the most recent data on the bottom), although this runs counter to the operation of most conventional strip chart recorders [82].

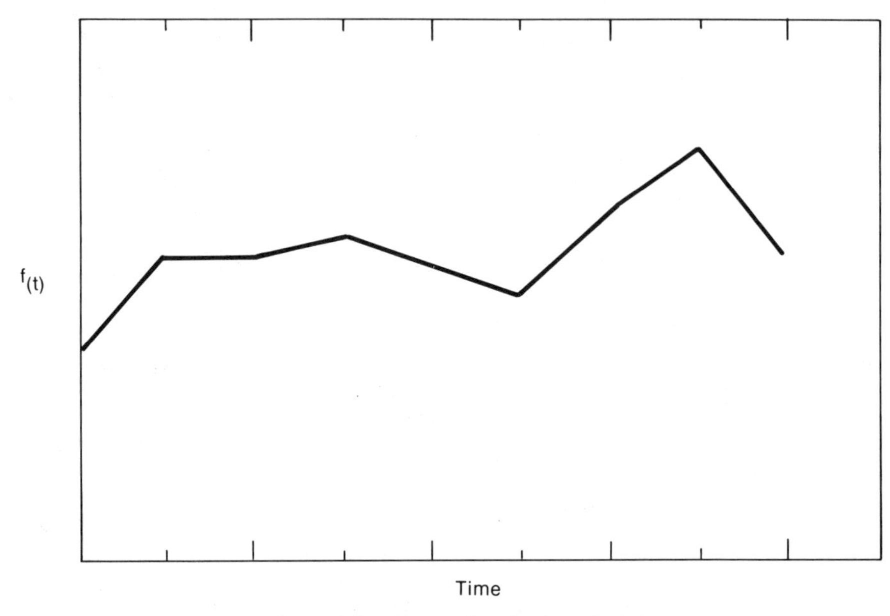

Figure 26. Example of a trend plot.

It is acceptable to display the minimum parameter set using several plots, each plot containing one or more variables. When more than one parameter is presented in a plot, there should be means of identifying each individual parameter. Color coding of traces is acceptable. Color codes used, however, must not conflict with other uses of color in the display [151].

Guidelines	Research Support	Source
1. If time is plotted on the X axis, it should increase from left to right; if time is plotted on the Y axis, it should increase from top to bottom.	Y	45, 82
2. When more than one parameter is presented in a plot, there should be means of identifying each individual parameter.	L	151

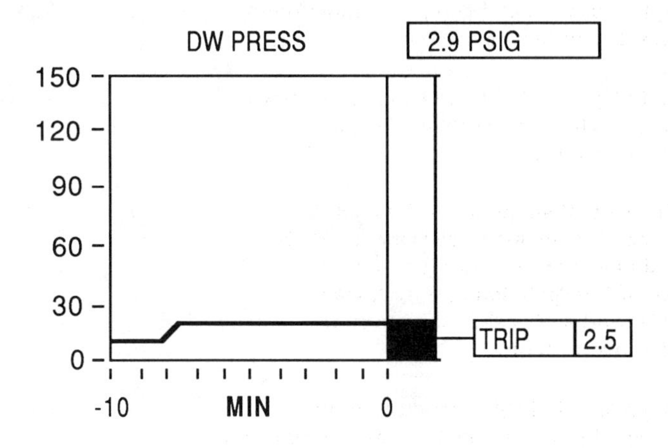

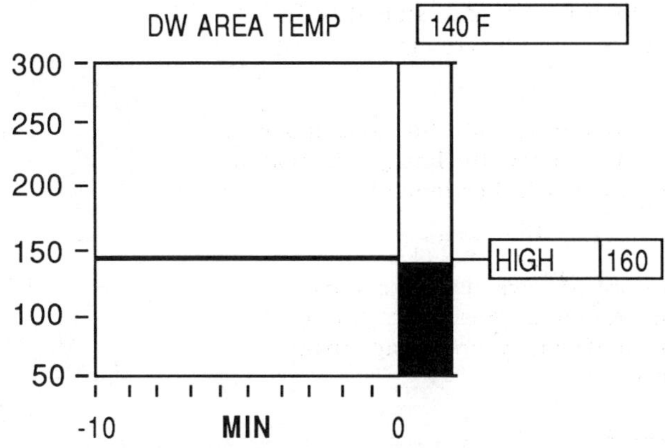

Figure 27. Special application of the trend plot. Bar charts and digital displays are used to supplement the information presented.

3.	When more than one parameter is displayed on a plot, the grouping of the parameters should enhance the operator's assessment of the safety status of the plant.	L	151, 39
4.	CRT-displayed trend plot scales should be consistent with the intended functional use of the data.[a]	L	150

5.	Graphic lines should contain a minimum of 50 resolution lines per inches.[b]	L	150
6.	Trend displays should be capable of showing data collected during time intervals of different lengths.[c]	L	39
7.	If the general shape of the function is important in making decisions, a graph should be chosen rather than a table or scale; if interpolations are necessary, graphs and scales should be used in preference to tables.	L	148
8.	Graphs should be constructed so that numbered grids are bolder than unnumbered grids. If ten-grid intervals are used, the fifth intermediate grid should be less bold than the unnumbered grids.	L	148
9.	For tasks requiring both time to estimate trends and accuracy, the line graph should be used rather than horizontal bar or column charts.	Y	126
10.	Time history displays of safety status parameters should present the 30-min interval immediately preceding current real time.[d]	L	151

---

[a] For example, the monitoring of neutron flux at reactor trip may have a variable scale of 0 to 100% of the design value and a time scale resolution of seconds. However, post-trip monitoring may have a variable scale of 0 to 10% with a time scale resolution of minutes. Finally, operational log data of neutron flux may have a time scale resolution of hours [150].

[b] A graphic line will appear continuous if the separation between addressable points, or resolution elements, is less than one minute of arc. To provide the illusion of continuity, graphic lines should contain a minimum of 50 resolution elements per inch.

[c] Reference [39] states, "A useful trend display should be capable of showing data collected during time intervals of different lengths. For example, a short time base of just a few minutes is needed to study fast changing trends, while

while other trends may not show significant changes for several hours. Although several variable trends may be grouped on the same display, it is very difficult to put an entire process overview in a single trend display. Grouped variables should be related, so the operator can correlate changes in one variable with changes in other key variables."

[d] Time history displays of parameters over a recent time interval are a preferred means of displaying trend and rate-of-change data. A time history of each safety status parameter for the thirty minutes immediately preceding current real time is acceptable. For applications to nuclear process control: availability of time history data displays on either the primary SPDS display format or on a secondary SPDS display format is acceptable [151].

Comment. This information format is most applicable when the operator's task requires presentation of historical or trend data. However, the implementation of this format should not violate the following constraints identified by Danchak [35].

Not to be used:

- When there are relatively few plotted values in the series,
- When emphasis should be on changes in amounts rather than on movement,
- To emphasize differences between values or amounts on different data,
- When movement of data is extremely violent or irregular, and
- When presentation is for popular appeal.

Caution should also be advised not to group more than three parameter lines on a single X-Y axis. Multiple parameter lines on the same axis should also be discriminative by some form of highlighting (dashed lines, solid lines, colors, etc.).

Mimic Display

A typical mimic display used for process control systems consists of a combination of graphic and alphanumeric elements. The features of this format generally take on the form of standard piping and instrumentation diagrams (P&IDs). Mimics integrate system components into functionally oriented diagrams that reflect component relationships. Properly designed mimics should decrease the operators decision-making load [150].

Extraneous detail will create a more cluttered mimic and may detract from meaningful information. For example, do not display bends in pipes just because there are bends in the actual pipe, unless this conveys useful information. Elements of a system that are physically close to each other should be displayed physically close in a mimic. However, for the sake of clarity, one may wish to move elements around on a display. In this case, demarcation lines may be used to indicate that the elements belong together.

An example of this information format is shown in Figure 28.

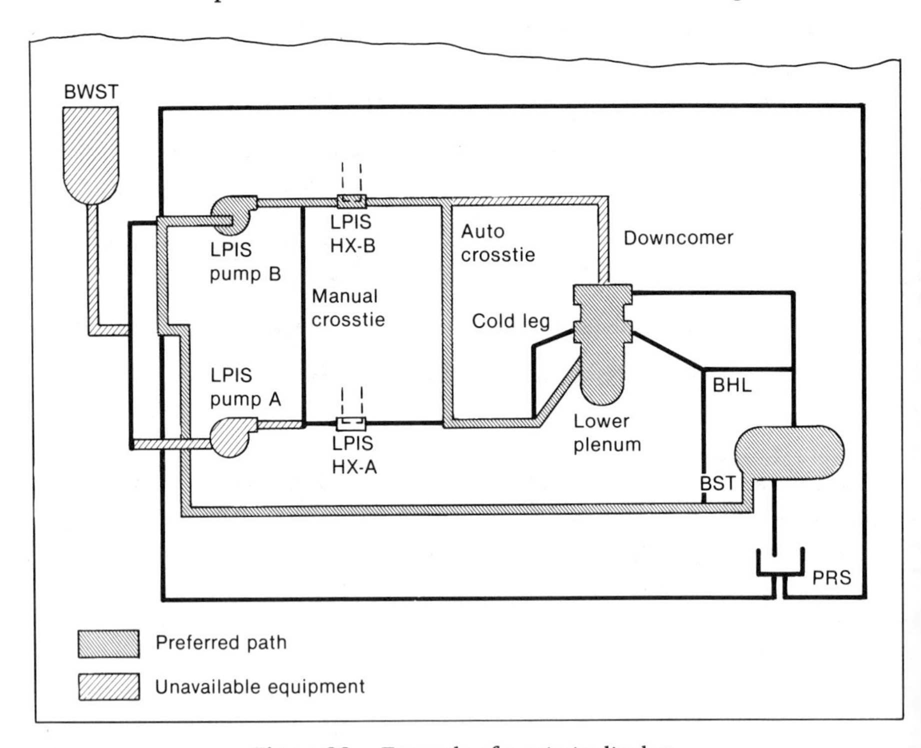

Figure 28. Example of a mimic display.

Guidelines	**Research Support**	**Source**
1. A mimic should contain just the minimum amount of detail required to yield a meaningful pictorial representation.	L	45
2. Abstract symbols should conform to common electrical and mechanical symbol conventions whenever possible.	L	150, 45
3. Differential line widths should be used to code flow paths (e.g., significance, volume, level).	L	150
4. Mimic lines should not overlap.	L	150
5. Flow directions should be clearly indicated by distictive arrowheads.		
6. All mimic origin points should be labeled or begin at labeled components.	L	150
7. All mimic destination or terminal points should be labeled or end at labeled components.	L	150
8. Component representations on mimic lines should be identified.	L	150
9. Symbols should be used consistently.	L	150

Comment. The uses of a dynamic mimic display to depict overall operational status and control of plant conditions can be extremely powerful. This technique aids the operator as an efficient device for overseeing the operation of various plant systems. Specific research data are limited, and detailed guidelines are dependent on the operator's tasks.

In addition, mimic displays are also made up of many components, and an effective mimic cannot be developed without attention to optimizing those other factors (color, graphic symbols, demarcation lines, labels, etc.).

Perhaps the greatest design shortcomings which may underutilize a mimic display's effectiveness are

- Screen clutter and layout–attempting to put too much information on the screen and/or adherence to detail beyond the needed requirements (i.e., screen clutter).
- Graphic symbols and labels–symbols (or labels) with intended meanings that are not meaningful or understandable and/or their placement is not in proximity to the elements they describe.
- Spatial relationships–the systems being mimicked do not conform to the spatial relationship of the actual process system.

VI

CONTROLS AND INPUT DEVICES

Keyboard Layout

The following are some of the most important characteristics of the keyboard:

- Keystroke feedback
- Key actuation force
- Key rollover
- Key travel (displacement)
- Key color/labeling
- Key dimension/spacing
- Keyboard slope
- Keyboard thickness
- Special function keys
- Soft programmable keys
- Numeric keypad

In the U.S., the basic alphanumeric set is comprised of 26 upper and lowercase characters, 10 numerals, and a special symbol set. In addition to the standard alphanumeric key set, the basic keyboard is usually accompanied with a set of function keys. The most common keys might be the return, shift, and reset. A special set of programmable function keys are also usually available. The cursor control key set consists of all devices for cursor positioning. This set encompasses the tab key and space bar. Many keyboards also possess the directional arrow cursor key set.

A standard QWERTY arranged keyboard, with the key groups previously described, is shown in Figure 29. For English-speaking countries, it might be noted that the QWERTY alphanumeric arrangement is the most commonly accepted standard. The other remaining key groups do not necessarily adhere to any known standardization.

This is also true of the special characters within the QWERTY arrangement (e.g., 0, 1, 7, %, etc.). The exception, in terms of numerical arrangement, might be the numerical key set. The selected locations for the function and cursor control key sets are highly varied (other than perhaps the space bar) and inconsistent amongst models. These differences cause initial confusion to the user when transitioning between keyboards of different models. There is an obvious need to standardize all keys, but, for the present, the current human factors data base is incomplete in establishing concrete guidelines.

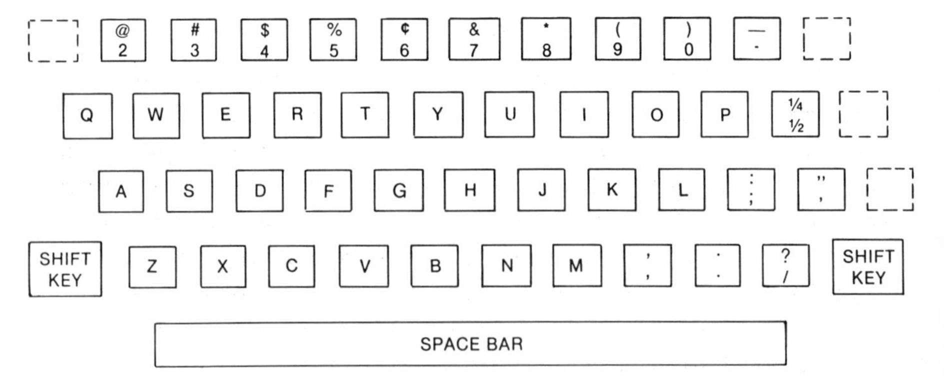

Figure 29. Standard QWERTY keyboard arrangement.

Keystroke Feedback

The subject of feedback in computer systems can take on many forms. Specific feedback to the operator can be tactile, auditory, or visual indication that the key has been activated. The assumption is made that the majority of keyboards display the correct symbol on command from the keyboard to the screen via the character generator logic. The presence of lag in the system from input (press key) to output (character presented on screen) is also assumed to be at a minimum. Therefore, with visual feedback interpreted as a given in most systems, the guidelines are pointed toward alternative modes of keystroke feedback (i.e., auditory, tactile).

The appropriateness of these forms are task- and context-dependent, but the literature indicates some general recommendations which are applicable to the majority of operating conditions.

The activation of each key should be accompanied by a feedback signal (e.g., audible click, tactile click, or snap action). Tactile feedback provided by a collapsing spring or a similar snap-action mechanism gives the key a positive feel which helps the operator to avoid missed or multiple keystrokes through uncertainty or as the result of teasing the key.

Guidelines	Research Support	Source
1. An indication of control activation should be provided (e.g., snap feel, audible click, or associated integral light).	L	150, 146, 22, 23, 46

Comment. The guidelines tend to overwhelmingly indicate that some form of keystroke feedback is a desirable feature. These guidelines are further substantiated from the authors' own informal observations. Data entry operators and programmers are generally opposed to keyboards with a mushy feel and prefer keyboards with a higher resistance. Evidence also indicates, for the novice or unskilled user, that keying is both faster and more accurate if tactile feedback is provided. However, tactile feedback seems to be less important for skilled operators (though still a desired attribute), and too positive a snap action may actually increase errors.

These recommendations for the merits of keystroke feedback as a function of operator skill may have directed keyboard designers to allow the operators the option of setting a switch to selectively choose the level of tactile/audible feedback desired. No distinction is implied in the current guidelines for one feedback mode (whether auditory or tactile) being superior over another. The choice is limited in situations where specialized membrane keyboards are used. Those devices generally require an auditory signal to compensate for the lack of tactile feedback typically encountered with standard equipment. It should be noted that this special requirement (auditory feedback only) would make the membrane keyboard only appropriate for extremely low-frequency keying tasks.

Key Actuation Force

The amount of resistance required to activate a key is usually expressed in Newtons, ounces, or grams. These guidelines are presented within an acceptable maximum and minimum range. In some situations, guidelines provided for numeric keys and alphabetic keys are different. The speed and accuracy with which a key is activated, as well as its susceptibility to inadvertent operation, are significantly affected by the type and amount of resistance built into the key. The ideal resistance level should produce the optimum combination of (a) precision, (b) speed, (c) control "feel," and (d) smoothness of movement. Selection of key actuation forces in early mechanical typewriters were severely restricted by design constraints. Modern keyboards are no longer subject to those early design constraints; as a result, greater flexibility is allowed for taking into account individual task and user requirements.

Guidelines	Research Support	Source
1. The force required for key displacement should be 0.25 to 1.5 N.[a]	Y	150, 22, 146, 23

2.	The force required for key displacement should be 0.3 to 0.75 N for repetitive keying tasks.	Y	42

[a] References [150] and [146] allow a range from 1 to 4 N for displacement of alphanumeric keys. Reference 42 specifies, for experienced operators, poke board membrane keyboard forces of about 0.6 N.

Comment. A small variance exists in the guidelines literature for key actuation force. In the face of this confusion, the guidelines from MILSTD 1472C [146[and NUREG-0700 [150] are recommended as the most acceptable standard to implement. This is due to their extensive previous applications in the design of military systems.

Key Rollover

A major characteristic which can help to minimize miskeying problems is key rollover. When keying at high speed, the user may produce bursts of keystrokes with very short interkeying times. That often results in more than one key being in the depressed state at a given time. The absence of key rollover results in lost keystrokes. With multiple-key rollover (often referred to as n-key rollover), the keyboard is able to store all the keystrokes and generate all the characters in their correct sequence regardless of the number of keys depressed at the same time. A graphical depiction of n-key rollover is shown in Figure 30.

	Guidelines	**Research Support**	**Source**
1.	N-key rollover capability should be implemented for the reduction of keying errors.	L	23

Comment. The experimental evidence is limited, but preliminary observations indicate that n-key rollover is highly preferred among operators. In addition, error rates have been shown to increase significantly when operators familiar with n-key rollover transfer to a keyboard without that capability.

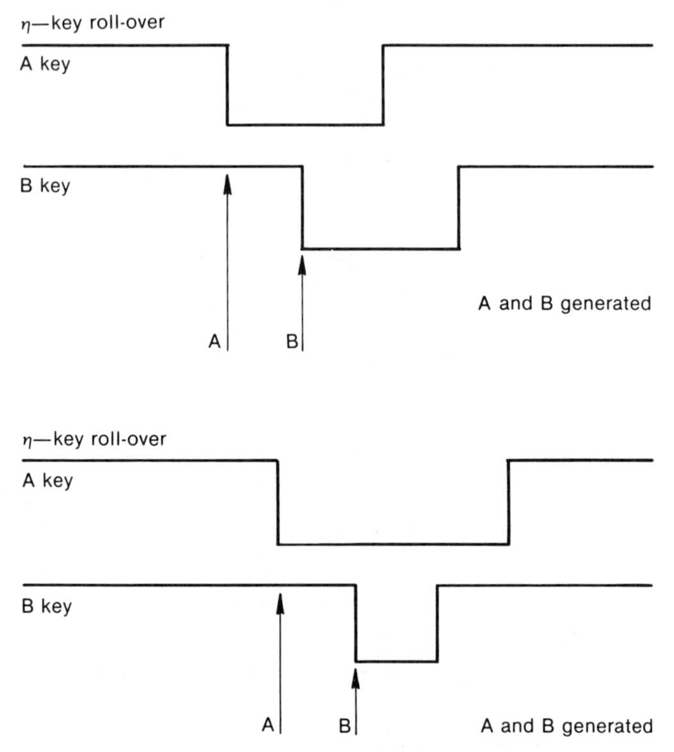

Figure 30. Examples of n-key rollover for two conditions: (a) depression of two keys when the first key is released before the second; (b) depression of two keys when the second key is released before the first key (Source: Reference [23]).

Key Travel (Displacement)

The actual distance the key travels in the vertical plane is the key travel or displacement. The distance is usually measured from the key's resting state to the activation point. This value is normally stated in inches or millimeters. These guidelines, as with the variable key actuation force, are presented within an acceptable maximum and minimum range. In some situations, separate guidelines are provided for both the numeric key set and the alphanumeric key set. In the same manner as key actuation force (resistance), prior constraints of earlier mechanical keyboards are no longer a major limitation for establishing key travel parameters. Rather, a greater flexibility is provided in modern keyboards for better taking into account individual task and user requirements. It might also be noted that optimal key travel is highly related to key actuation force.

Guidelines	Research Support	Source
1. Key displacement should be 0.03 to 0.19 in. for numeric keys and 0.05 to 0.25 in. for alphanumeric keys.[a]	Y	150, 146
2. Displacement variability between keys should be minimized.	Y	42

[a] Reference [23] recommends a key displacement of 0.8 to 8 mm. Reference [42] recommends a key displacement of 1.3 to 6.4 mm for a standard key and even less with poke board or touch pad. Reference [69] recommends a key displacement of 1.3 to 6.4 mm. Reference [22] recommends a key displacement of 5 to 8 mm. Reference [5] recommends a key displacement of 0.125 to 0.187 in. Reference [42] recommends an alphabetic key displacement of 1/16 to 1/4 in. and a numeric key displacement of 1/32 to 3/16 in.

Comment. A small variance exists in the literature for key travel. This situation is similar to the guidelines for key actuation force. It is recommended, due to extensive previous applications in design of military systems, that the guidelines from MIL-STD-1472C [146] and NUREG-0700 [150] be implemented. It is also suggested that, due to the highly interrelated nature of key force with displacement, high force values not be implemented with low displacement. Preliminary research indicates this force-displacement relationship was most disliked and difficult to use.

In conclusion, little reference to criteria for membrane keyboards was cited. Eastman Kodak [42] recommended the values for displacement should be less than that for a standard keyboard, but no actual data are available. It is not clear from Cakir et al. [23] that their recommendation (0.8 to 8 mm) did not also include specification for a membrane keyboard at the minimum value. It is highly unlikely that 0.8 mm is sufficient for a standard keyboard arrangement.

Key Color/Labeling

The use of color for key caps satisfies a multiple purpose in keyboard layout: (a) color coding of functional groups, (b) background to provide sufficient contrast for key labels, and (c) glare reduction. In a typical touch-typing task, the operators seldom rely on key labels, and their function is considered secondary. However, the addition of many specialized function and control keys on modern complex keyboards have created a necessity for good labeling and key coloring techniques for the novice as well as the skilled typist. Even experienced touch typists must occasionally scan the keyboard for specialized keys not within the standard QWERTY arrangement. As a result, the labels, and colors incorporated into the keys take on a new importance as an operator aid. These parameters (key color and labels) will be treated as a single variable due to their highly interrelated nature.

	Guidelines	**Research Support**	**Source**
1.	All controls should be appropriately and clearly labeled in the simplest and most direct manner possible.	L	5
2.	Functional highlighting of the various key groups should be accomplished through the use of color-coding techniques.	L	5
3.	Key symbols should be etched to resist wear and colored with high contrast lettering.	L	23
4.	Color of alphanumeric keys should be neutral (e.g., beige, grey) rather than black or white or one of the spectral colors (red, yellow, green, or blue.)	L	23
5.	Keys should be matt finished.	L	Authors
6.	Keys should be labeled with a nonstylized font.	L	Authors

Comment. The information is scarce, but the evidence indicates that matt-finished keys are most preferable and neutral colors are the most acceptable key colors for minimization of glare. In summary, the authors conclude from the literature that key labels, and colors should conform to the following criteria:

- Keys are matt finished.
- Key symbols are etched (molded into the key top) to resist wear and are of suitable contrast for legibility under a wide range of lighting conditions.
- Keys are color coded by functional groups with a neutral color scheme which is readily discriminable and of sufficient glare reducing quality.
- Key labels are simple and direct in a basic, nonstylized font (see also section on "alphanumeric characters").

Key Dimension/Spacing

The size and spacing of keys are directly related to the keyboard operator's performance. Optimal dimensions and spacing are fundamental to keyboard usability. The size of the key top is to some extent the result of a compromise between providing a large enough target area and surface for the legend and keeping the keyboard dimensions and distance of finger travel down to a manageable size. Key dimension/spacing are discussed as a single variable due to their highly functional relationship. Key spacing is generally described as the distance between key centers. Generally, key tops should be square or slightly rounded. Other specifications address key spacing as the distance between adjacent edges.

	Guidelines	Research Support	Source
1.	The linear dimensions of the key tops should be from 0.385 to 0.75 in., with 0.5 in. preferred.[a]	L	150, 5, 146, 67 42, 23,46
2.	Separation between adjacent key tops should be 0.25 inch.[b]	L	150, 5, 146, 67, 42, 23,

3.	Push-button height for decimal entry key pads should be from 1/4 to 3/8 inch.	L	46
4.	Key height for alphanumeric keyboards should be from 3/8 to 1/2 inch.	L	46

[a] Reference [23] recommends 12 to 15 mm. Reference [42] recommends 12 mm square and notes that miniature key sets are associated with decreased speed and increased errors. Reference [46] specifies from 3/8 to 1/2 in.

[b] Reference [42] recommends 19 mm between key centers. Reference [23] recommends from 18 to 20 mm between key centers.

Comment. The authors conclude that the guidelines from MIL-STD-1472C [146] and NUREG-0700 [150] are the accepted standard due to their high application and proven performance in military systems. Key tops should be square or slightly rounded with a diameter of 0.5 in. Square tops are generally regarded to be more suitable than round since they provide a bigger target area between key centers. Separation between adjacent edges should be 0.25 in. (center to center spacing, 0.75 in.). Key tops should also be concave.

In conclusion, it is not apparent why Reference [46] made a distinction for key height between numeric and alphanumeric keys. The authors suggest that key height should be dependent on the key displacement provided.

Keyboard Slope

Keyboard Slope is expressed as the angle of the keyboard relative to the horizontal surface. An example of this angle is illustrated in Figure 31. This value is usually expressed as an acceptable range in units of degrees. The angle of the keyboard is important from the point of view of the position and of the hands and fingers while keying.

	Guidelines	**Research Support**	**Source**
1.	Keyboards should have a slope of 15 to 25 degrees from the horizontal, with 12 to 18 degrees preferred.[a]	Y	150, 5, 146, 67, 23, 22, 47

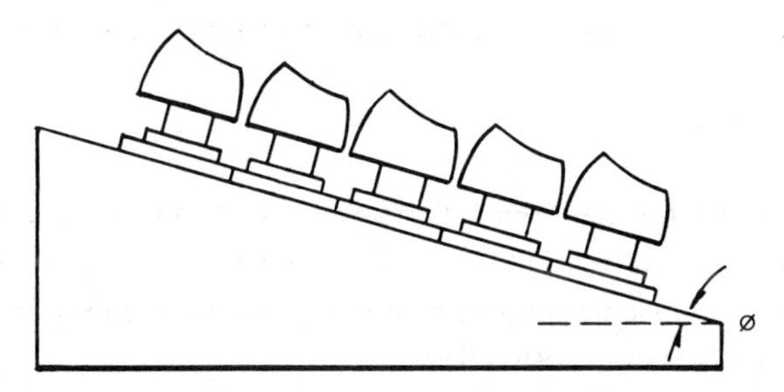

Figure 31. Keyboard angle (θ) relative to the horizontal surface it sits on.

2. The keyboard slope should be adjustable. Y 7

[a] Reference [5] recommends 11 to 15 degrees. Reference [67] recommends 10 to 35 degrees. Reference [47] recommends 15 to 20 degrees. Reference [23] recommends 5 to 15 degrees. Reference [22] recommends less than or equal to 5 degrees.

Comment. A summary of related keyboard studies is described below:

- Alden [3] indicates that keying performance was found to be remarkably stable over a wide range of keyboard slopes.
- Studies show that operators prefer some slope to the keyboard and that, as a result, there is a wide range of individual preference.

This rationale implies the adjustable keyboard to be the most desired as recommended by Reference [7]. Cakir et al. [22], in addition to keyboard slope, also make reference to criteria for keyboard profile. Stepped, sloped, and dished were three profiles identified. It is not evident that those factors significantly affect human performance when compared with keyboard slope.

Most of the above guidelines are in general agreement for acceptable ranges. The guidelines from Reference [22] of less than 5 degrees possessed the most deviation from this group. The attempt to minimize keyboard height is the apparent reason for the unusual keyboard slope in that report. The author concludes that if the keyboard is not adjustable, the guidelines from MIL-STD-1472C [146] and NUREG-0700 [150] are the most accepted standard due to their high application and proven performance in military systems.

Keyboard Thickness

The value for keyboard thickness is described as the vertical height of the keyboard from its base to the top of the home row keys. This variable is most directly relevant to postural loading of the user by ensuring the correct working level.

Guidelines	Research Support	Source
1. The thickness of the keyboard, i.e., base to the home row of keys, should be less than 50 mm (acceptable) with 30 mm or less preferred.[a]	L	23, 22, 41

[a] Reference [22] recommends that keyboards more than 30 mm thick be recessed into the table top.

Comment. This variable is highly dependent on the overall layout of the workstation. These guidelines imply that thinner is better.

Reference [22] justifies this requirement in order to meet the needs of a suitable working level. Working level is defined as the distance from the underside of the thighs and the palms of the hand (see Figure 32). This value should fall in the range from 220 to 250 mm (8.7 to 9.8 inches).

Therefore, it is safe to state that a keyboard thickness of less than 30 mm is desired in order to achieve an optimal working level.

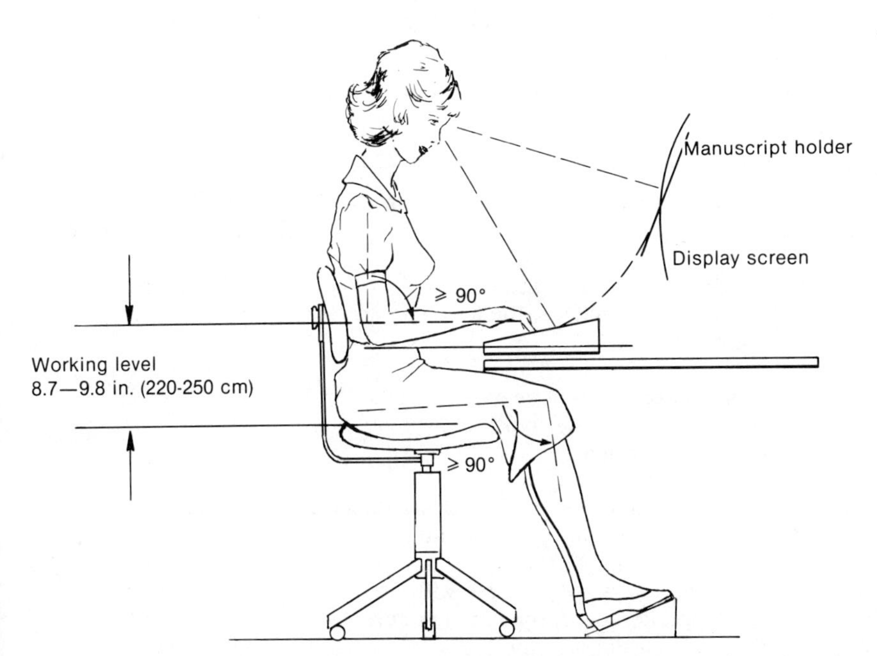

Figure 32. The ideal working level is defined as the distance between the underside of the thighs and the palms of the hands when operating the keyboard with the hands and forearms in an approximately horizontal position (Source: Adapted from Reference [23]).

(See also the guidelines for chair design.) The merits of a thinner keyboard will also allow the operator to utilize the desk top as a convenient palm rest if it is not readily provided on the keyboard.

Special Function Keys

In addition to the standard alphanumeric keys, a typical keyboard often consists of a special function key set. These keys encompass a wide range of control capability. They can be used to perform routine typewriter functions or as a device to select menus on the screen. This latter function usually involves the use of programmable function keys as opposed to the dedicated hardwired function keys.

It is important that the layout of the keys is such that typical operational sequences form a logical flow on the keyboard. This helps to reduce errors and to maintain the operators' keying rhythm. However, it is also important that the layout minimizes the effect of likely errors. Overreaching frequently used keys or failing to change shift are both extremely common and should not result in critical errors.

Guidelines	**Research Support**	**Source**
1. When dedicated controls are used to initiate/activate functions, the keys should be grouped together.	L	150
2. Function controls should be easily distinguished from other types of keys on the computer console.	L	150
3. Each function control should be clearly labeled to indicate its function to the operator.	L	150
4. When function keys are included with an alphanumeric keyboard, the function keys should be physically separate.[a]	L	150
5. Keys with major or fatal effects should be located so that inadvertent operation is unlikely.	L	23

[a] Avoid multiple-mode keyboards which utilize the same keys for both alphanumerics and functions by using shift keys or mode selection controls [150].

Comment. The guidelines listed above are extremely general in that their origins reside in standard human engineering practices. In the majority of situations, the establishment of specific guidelines are dependent upon the user's task requirements. In spite of these limitations, some general statements can be gleaned from the above items:

- Function Keys should be grouped on the keyboard according to one or more of the following techniques: (a) sequence, (b) frequency, (c) function, or (d) importance.
- Function Keys should be distinct from the other keys by (a) color, (b) shape, and/or (c) spacing.
- If more than one keyboard exists in the control room, design and layout of the function keys should be consistent for all keyboards.
- Avoid multiple-mode keyboards that include functions and alphanumeric characters on the same keys which are controlled with a shift key or mode select switch. (function keys should be physically separate.)

Soft Programmable Keys

Soft programmable keys, or programmable display push buttons (PDPs) as they are called in the space transportation industry [156], have a high degree of flexibility and are a relatively recent innovation. The type used to date relies on software to turn on different LED's in a pattern relative to the task at hand. PDPs can be used to help users step through a menu hierarchy or checklist [156]. Because software changes the face and function of the button, the number of displays or switches a button is capable of is limited only by the software and nature of the mission at hand.

Guidelines	**Research Support**	**Source**
1. Commands should be consistent throughout PDP procedures.	L	156
2. Use blink coding when there is an urgent need for the subject's attention.	L	156, 135
3. The system should allow users to step backward or forward through menus or procedures.	L	156
4. PDPs should not be used in complex applications such as the sole display and control, e.g., use in conjunction with CRT.	L	156
5. PDPs may be used as the sole device with simple applications such as camera control.	L	156
6. PDPs should contain abbreviations which are easily recognized by the user. (In many cases there is a six-letter limit on a button for labels.)	L	156

Numeric Keypad

The numeric keypad is an optional or auxiliary device to the standard keyboard. The keys (0-9) are generally laid out in a hand-held calculator arrangement. That is, a 3 x 3 matrix with 7, 8, 9 on the top row, 4, 5, 6 on the second row, and so forth. The "0" key is usually reserved on a separate row immediately below the 3 x 3 matrix. The numeric keypad provides a special capability when rapid entry of large amounts of numerical data is required. In general, the numeric pad operates in precisely the same way as the numeric keys in the main alphanumeric key set except it is not affected by the shift keys. The numeric keypad is highly beneficial if the task requires a substantial entry of numbers not mixed into text.

Reference [42] reports, "Redundancy of the numeric keys (across the top of the keyboard and also as a separate numeric set next to the typewriter key set) can be helpful if a terminal may be used as a calculator as well as an alphanumeric data entry keyboard." The layout of the numeric keys within the keyboard is highly debatable, but most researchers agree that the configuration should conform to a 3 x 3 matrix arrangement with "0" at the bottom row. However, some general disagreement exists in the organization of the keys within the matrix. This narrows down to two basic arrangements: (a) the touch tone telephone pad and (b) the calculator or adding machine layout. Examples of these two schemes are shown in Figure 33.

	Guidelines	Research Support	Source
1.	Terminals which are often used as calculators should be provided with an auxiliary numeric key set.	Y	42
2.	The configuration of a keyboard used to enter solely numeric information should be a 3 x 3 x 1 matrix with the zero digit centered on the bottom row.	Y	150, 146, 47
3.	The layout of keyboard numeric pads should be either telephone or calculator style.[a]	Y	150, 5, 67, 23, 47, 46

[a] Reference [23] notes that push-button telephones are increasingly being used in conjunction with VDUs. References [5], [67], [23], [47], and [46] state preference for the telephone style.

Comment. It is interesting to note that many of the guidelines emphasize a specification for keyboard arrangement, i.e., telephone or calculator. Some of the guidelines cite a preference, while NUREG-0700 [150] states that either is acceptable. Eastman Kodak [42] also leaves the preference optional to the designer. MIL-STD-1472C [146] only recommends the configuration (3 x 3). Cakir et al. [23] defends the telephone arrangement since many users usually time-share their tasks across both devices. That would be an acceptable rationale for recommending the telephone arrangement in a control room. That logic, unfortunately, breaks down, since many operators perform routine calculations on a hand-held calculator as well as talk on the

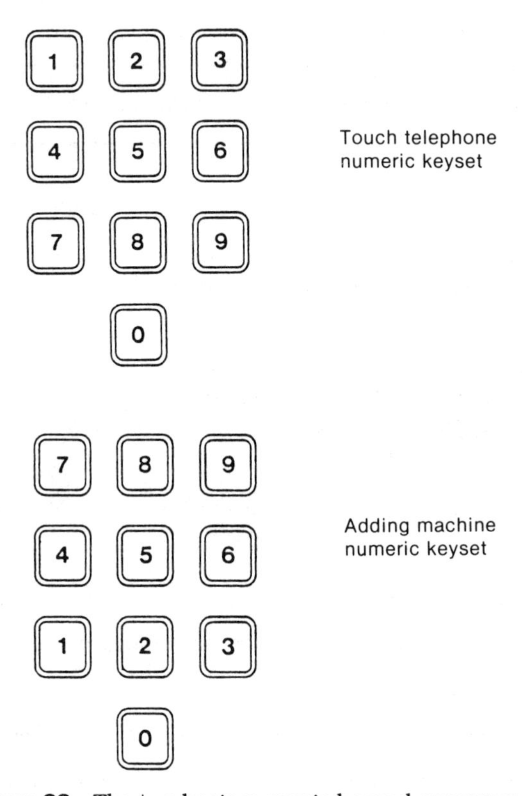

Figure 33. The two basic numeric keypad arrangements.

phone. The argument is also certainly moot for most process control applications, since high-speed entry of data is usually not required of operators.

In summary, the research indicates that a slight advantage can be obtained using the telephone arrangement. For applications where a premium is not placed on rapid entry of numerical data, either the calculator or telephone arrangement in a 3 x 3 x 1 matrix is acceptable [150].

Alternate Input Devices

In addition to the keyboard, a variety of other mechanisms can be used to regulate or guide the operation of a machine, apparatus, or system and are discussed in this section.

- Light pens (see Figure 34)
- Joysticks (see Figure 35)
- Tracker ball (see Figure 36)
- Grid and stylus devices (see Figure 37)
- Free-moving X-Y controller (mouse) (see Figure 38)
- Automatic speech recognition (voice input device)
- Touch screen input

There are six items to consider in the selection of an alternate control type:

1. What type of control input is required by the CRT (discrete state or continuous range)?
2. If discrete changes are needed, how many states are required? If continuous range is appropriate, what is the value, precision, force, and speed of movement required?
3. Is there a danger of confusing controls?
4. What feedback is there that the system has accepted the control input?
5. How can unintentional control operation (bumping) be reduced?
6. What population stereotypes relative to control movements, shape, and size are considered natural and compatible?

Light Pens

Light pens have been used with some degree of success. Those with connector cables do pose some risk, as cables can be broken and pens can be lost. In critical conditions, they may not be as desirable as some more rugged input devices, such as the tracker ball.

Guidelines	Research Support	Source
1. Light pens should be used for cursor placement, text selection, and command construction.[a]	Y	146, 119, 120
2. Tasks involving light pens should not require frequent, alternating use of the light pen and the keyboard.[b]	Y	48, 57, 77
3. Tasks involving light pens should not require long, continuous intervals of light pen use.[c]	Y	48, 57, 77

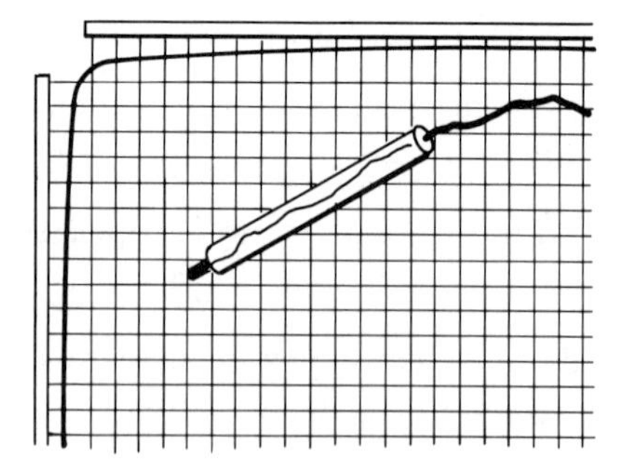

Figure 34. Light pen.

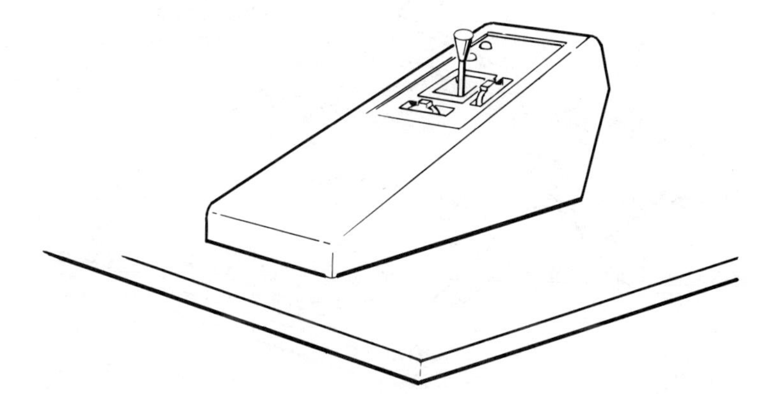

Figure 35. Joystick.

Figure 36. Tracker ball control.

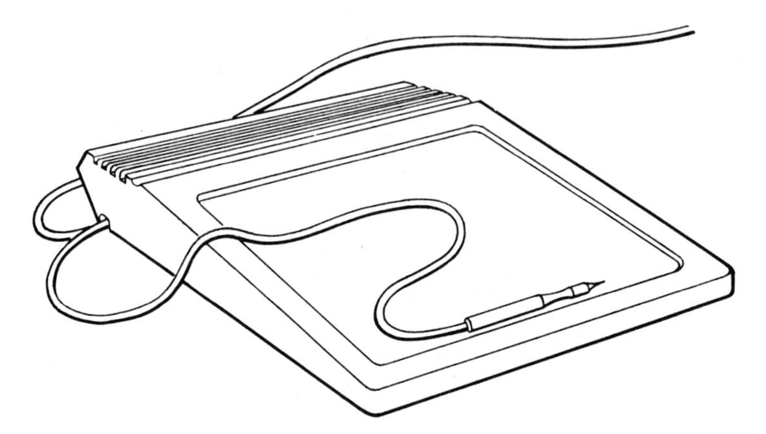

Figure 37. Grid and stylus devices.

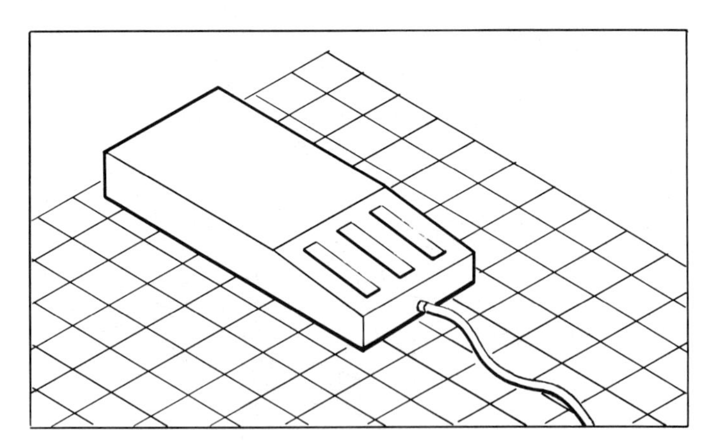

Figure 38. Free-moving X-Y controller (mouse).

4.	The light pen should be 12 to 18 cm (4.7 to 5.1 in.) long and 0.7 to 2 cm (0.3 to 0.8 in.) in diameter.	Y	147, 45
5.	Convenient clips should be provided at the lower right side of the CRT to hold the pen when it is not in use.	Y	146
6.	Movement of the pen in any direction on the screen should result in smooth movement of the follower in the same direction.	Y	146
7.	Discrete placement of the stylus at any point on the screen should cause the follower to appear at that point and remain steady in position so long as the pen is not moved.	Y	146
8.	Refresh rate for the follower should be sufficiently high to ensure the appearance of a continuous track whenever the pen is used for generation of free-drawn graphics.	Y	146

[a] Reference [146] recommends that a simple light pen may be used as a track-oriented readout device. That is, it may be positioned on the display screen to detect the presence of a computer-generated track by sensing its refresh pattern; the display system will then present a follower (hook) on the designed track. With suitable additional circuitry, a follower can be made to track the movement of the light pen across the surface, thus allowing it to function as a two-axis controller capable of serving the same purposes as the grid and stylus devices.
References [48], [57], and [77] state that light pens can be used effectively for cursor placement and text selection, command construction, and for interactive graphical dialogues in general, including drawing. However, there is evidence that greater accuracy may be possible with a mouse, in discrete tasks and with a tracker ball in drawing tasks.
[b] Mode mixing, as by alternating use of light pen and keyboard, can significantly disrupt performance, since the light pen must be picked up and replaced with each interval of use [48], [57], [77].

[c] Continuous use of a light pen, at least on commercially available CRT terminals with vertical display surfaces, can be quite fatiguing [48], [57], [77].

Joysticks

Reference [120] recounts, "There are many studies of the use of joysticks for continuous tracking tasks but few studies of its use for discrete or continuous operand selection or graphical input tasks. Those studies which have been performed have found the mouse, light pen, and tracker ball preferable in terms of speed, accuracy, or both. Joysticks are sometimes used for windowing and zooming control in graphical displays. No research on this topic was found, although the results of tracking studies may be applicable here. Otherwise, no clear recommendations for joystick properties emerged from the survey, even with respect to basic issues like position vs. rate vs. acceleration control. These issues may be fairly task-specific."

Guidelines for joysticks are as follows:

Guidelines	Research Support	Source
1. Joystick controls should be used for tasks that require precise or continuous control in two or more related dimensions.	Y	146
2. In rate-control applications which allow the follower to transit beyond the edge of the display, indicators should be provided to aid the operator in bringing the follower back onto the display.	Y	146

Isotonic Joysticks. Isotonic joysticks are free-moving controls with feedback coming from the distance moved. Guidelines for isotonic joysticks are as follows:

Guidelines	Research Support	Source
1. Isotonic joysticks which are used for rate control should be spring-loaded for return to center when the hand is removed.	Y	146
2. Isotonic joysticks should not be used in connection with automatic sequencing of a CRT follower unless they are instrumented for null return or are zero-set to the instantaneous position of the stick at the time of sequencing. Upon termination of the automatic	Y	146

	sequencing routine, joystick center should again be registered to scope center.		
3.	Isotonic/displacement joysticks should be 1/4 to 5/8 in. in diameter and 3 to 6 in. long.	Y	146, 45
4.	Resistance force of the joystick should be 12 to 32 ounces.	Y	146, 45
5.	Full displacement of the joystick should not exceed 45 degrees.	Y	146, 45
6.	Isotonic/displacement joysticks should be provided with the following clearances: a. Display to stick–15-3/4 in. b. Around stick–4 in. c. Stick to shelf front–4-3/4 in. to 9-7/8 in.	Y	146, 45
7.	Movement should be smooth in all directions, and rapid positioning of the follower on the display should be attainable without noticeable backlash, cross-coupling, or need for multiple corrective movements.	Y	146
8.	Control ratios, friction, and inertia should meet the dual requirements of rapid gross positioning and precise fine positioning.	Y	146
9.	Recessed mounting or pencil attachments may be utilized to provide greater precision of control.	Y	146
10.	When used for generation of free-drawn graphics, the refresh rate for the follower on the CRT should be sufficiently high to ensure the appearance of a continuous track.	Y	146
11.	Delay between control movement and the confirming display response should be minimized and should not exceed 0.1 s.	Y	146
12.	When positioning accuracy is more critical than positioning speed,	Y	146

	isotonic displacement joysticks should be selected over isometric joysticks.		
13.	Isotonic displacement joysticks should be used for such functions as data pickoff and generation of free-drawn graphics.	Y	146

Isometric Joystick. An isometric joystick (also known as stiff stick, force stick, or pressure joystick) is a lever that doesn't move; output exists as a function of applied force. Isometric joysticks are particularly appropriate for applications: (a) which require return to center after each entry or readout; (b) in which operator feedback is primarily visual from some system response rather than kinesthetic from the stick itself; and (c) where there is minimal delay and tight coupling between control and input and system reaction [146].

Guidelines for isometric joysticks are as follows:

	Guidelines	**Research Support**	**Source**
1.	The isometric joystick should be used for such functions as data pickoff.	Y	146
2.	Isometric joysticks should ordinarily not be used in any application where it would be necessary for the operator to maintain a constant force on the stick to generate a constant output over a sustained period of time.	Y	146
3.	Finger-grasped isometric joysticks should comply with the same dimensional criteria as isotonic joysticks.	Y	146
4.	Hand-grasped isometric joysticks, when integral switching is required, should be between 4.3 to 7.1 in. long and have a maximum grip diameter of 2 in.	Y	146
5.	Hand-grasped isometric joysticks should have minimum clearances of 4 in. at the sides and 2 in. at the rear.	Y	146

6.	Hand-grasped isometric joysticks should have a maximum resistance force of 26.7 lb for full output.	Y	146
7.	The isometric stick should deflect minimally in response to applied force but may deflect perceptibly against a stop at full applied force.	Y	146
8.	The X and Y output should be proportional to the magnitude of the applied force as perceived by the operator.	Y	146

Tracker Ball

The tracker ball control does not provide an automatic return to point of origin; hence, if used in applications requiring automatic return to origin, following an entry or readout, the interfacing system must provide this. Because the ball can be rotated without limit in any direction, it is well suited for applications where there may be accumulative travel in a given direction. Tracker ball controls should be used only as position controls (i.e., a given movement of a ball makes a proportional movement of the follower on the display) [146]. They should be considered in instances where there is inadequate desk or table top space for a mouse.

Guidelines for tracker ball controls are as follows:

	Guidelines	**Research Support**	**Source**
1.	A ball control should be used for such tasks as data pickoff.	Y	146
2.	In any application with tracker ball controls, do not allow the ball to drive the follower on the display off the edge of the display.	Y	146
3.	When tracker ball controls are used to make precise or continuous adjustments, wrist support or arm support or both should be provided.	Y	146

4.	Tracker ball controls should conform to the dimensions listed in Table 11.	Y	146, 45
5.	The tracker ball control should be capable of rotation in any direction so as to generate any combination of X and Y output values.	Y	146
6.	When moved in either the X or Y directions alone, there should be no apparent cross-coupling (follower movement in the orthogonal direction).	Y	146
7.	While manipulating the control, neither backlash nor cross-coupling should be apparent to the operator.	Y	146
8.	Control ratios and dynamic features should meet the dual requirement of rapid press positioning and smooth, precise fine positioning.	Y	146
9.	Tracker balls should be used in graphic applications requiring position and selection.	L	68

Grid and Stylus Devices

Reference [146] states that grid and stylus devices may be used for data pickoff from a CRT, entry of points on a display, generation of free-drawn graphics, and similar control applications. The grid may be on a transparent medium, allowing stylus placement directly over corresponding points on the display, or it may be displaced from the display in a convenient position for stylus manipulation. In either case, a follower (bug, mark, hook, etc.) should be presented on the display at the coordinate values selected by the stylus. Devices of this type should be used only for zero-order control functions (i.e., displacement of the stylus from the reference position causes a proportional displacement of the follower). Reference [48] states that graphical input tablets are capable of fairly high pointing accuracy (within 0.08 cm, according to one study). They are commonly used for freehand drawing but may be inferior for discrete position input tasks. They may also involve a performance decrement due to low stimulus-response compatibility when the drawing surface is separate from the display surface.

Table 11. Minimum, maximum, and preferred dimensions of ball controls. (Source: Reference [146])

	DIMENSIONS		RESISTANCE		CLEARANCE		
	D DIAM	A SURFACE EXPOSURE	PRECISION REQUIRED	VIBRATION OR ACCEL CONDITIONS	S DISPLAY CL TO BALL CL	C AROUND BALL	F FALL TO SHELF FRONT
MINIMUM	50 mm (2 in.)	1545 mrad (100°)			0	50 mm (2 in.)	120 mm (4-3/4 in.)
MAXIMUM	150 mm (6 in.)	2445 mrad (140°)	1.0 N (3.6 oz)	1.7 N (6 oz)	320 mm (12-5/8 in.)		250 mm (9-3/4 in.)
PREFERRED	100 mm (4 in.)	2095 mrad (120°)	0.3 N (1.1 oz)				

Guidelines for grid and stylus devices are as follows:

Guidelines	Research Support	Source
1. Grid and stylus devices should be used for data pickoff, entry of points on a display, generation of free-drawn graphics, and similar control applications.	Y	146, 48
2. Transparent grids which are used as display overlays should conform to the size of the display.	Y	146
3. Grids which are displaced from the display should approximate the display size and should be mounted below the display in an orientation to preserve directional relationships to the maximum extent (i.e., a vertical plane passing through the north/south axis on the grid	Y	146

	shall pass through or be parallel to the north/south axis on the display).		
4.	Movement of the stylus in any direction on the grid surface should result in smooth movement of the follower in the same direction.	Y	146
5.	Discrete placement of the stylus at any point on the grid should cause the follower to appear at the corresponding coordinates and to remain steady in position so long as the stylus is not moved.	Y	146
6.	Refresh rate for the follower should be sufficiently high to ensure the appearance of a continuous track whenever the stylus is used in generation of free-drawn graphics.	Y	146

Free-Moving X-Y Controller (mouse)

This type of controller may be used on any flat surface to generate X and Y coordinate values which control the position of the follower on the associated display. It may be used for data pickoff or for entry of coordinate values. It should be used for zero order control only (i.e., generation of X and Y outputs by the controller results in proportional displacement of the follower). It should not be used for generation of free-drawn graphics [146].

Reference [146] recommends the following dimensions:

	Minimum	**Maximum**
Width (*spanned by thumb to finger grasp*)	40 mm (1.6 in.)	70 mm (2.8 in.)
Length	70 mm (2.8 in.)	120 mm (4.7 in.)
Thickness	25 mm (1.0 in.)	40 mm (1.6 in.)

When the operator grasps the controller in what seems to be the correct orientation; and moves it rectilinearly along what is assumed to be straight up the Y axis, then the direction of movement of the follower on the CRT shall be between 6110 and 175 mrad (350° and 10°). The controller shall be easily movable in any direction without a change of hand grasp and shall result in smooth movement of the follower in the same direction ±175 mrad (10°).

Guidelines for the free-moving X-Y controller (mouse) are as follows:

Guidelines	Research Support	Source
1. The mouse controller should be used for main item selection, scrolling, data retrieval, and data entry.	Y	146
2. The controller should have physical dimensions of 1.5 to 3 in. width, 3 to 5 in. length, and 1 to 2 in. thickness.	Y	146, 45
3. The design of the controller and placement of the maneuvering surface should allow the operator to consistently orient the controller to within ±175 mrad (10°) of the correct orientation; without visual reference to the controller.	Y	146
4. The controller should be easily movable in any direction without a change of hand grasp and should result in smooth movement of the follower in the same direction ±175 mrad (10°).	Y	146
5. The controller should be cordless and should be operable with either the left or right hand.	Y	146
6. A complete excursion of the controller from side to side of the maneuvering area should move the follower from side to side on the display regardless of scale setting or offset unless expanded movement is selected for an automatic sequencing mode of operation.	Y	146
7. Applications which allow the controller to drive the follower off the edge of the display should provide indicators to assist the operator in bringing the follower back onto the display.	Y	146

Automatic Speech Recognition (Voice input device)

The current state of this technology limits its use to relatively simple input tasks. Even there, there are problems with different speakers, noise, etc. Although speech input seems like a very desirable and natural input mode and is clearly preferred over other communication modes for interpersonal communication, it is not clear whether it will prove to be widely applicable for human-computer interaction tasks. Very little information was found that would assist the designer in recognizing tasks for which speech input is appropriate or in selecting an appropriate speech input device [120]. As a strictly cultural phenomenon, languages such as Japanese and Chinese, where written characters are long and complex, may be more suitable for voice input applications in process control. Research indicates that voice is also poor for indicating locations [156]. For example, it is particularly poor for positioning cursors.

	Guidelines	Research Support	Source
1.	Automatic speech recognition (voice input devices) should be limited to relatively simple input tasks.	Y	120
2.	Keyboard is to be preferred for entry of numeric strings.	Y	156
3.	Voice entry may have an acceptable error rate for entry of alphanumeric strings.	Y	156

Touch Screen Input

Very early results with touch screens were less than satisfactory and the technology has, thankfully, responded. Reference [84] states, "These offer the most natural action of all: You simply point with your finger at something on the screen."

"Touch screens are the easiest pointing devices to use. They work well for novices and in situations where you are simply passing through–at a store display, for instance. Most let you perform menu selections, something you can do more quickly from the keyboard with a little experience."

Reference [7] narrates, "No standards reviewed by these authors contained any useful guidelines regarding the utility, application, or specifications for touch panels. They are a valid input medium, but research is needed to determine the optimal amount of screen area needed for presentation, errors in activations, etc."

Guidelines for touch screens are as follows:

Guidelines	Research Support	Source
1. Touch screens should be used for main item selection, scrolling, data retrieval, and data entry.	Y	107
2. The terminal should recognize a person's touch in approximately 100 ms.	Y	107
3. The system should accept only one command at a time, indicate that the command has been accepted, and respond in a time commensurate with the activity.	Y	107
4. The sensitive areas should be large enough to allow entry using fingers and allow for parallax due to CRT screen curvature.[a]	Y	107
5. To avoid alteration of color codes, touch screens should be toned with a neutral tint.	L	Authors
6. Touch screens are not recommended if task requires holding arm up to the screen for long periods of time.	L	Authors
7. Use discriminable audible beeps (used to supply feedback) when more than one touch screen will be installed at more than one workstation.	L	Authors

[a] To minimize parallax, touch screens that mount flush with the CRT screen, or as close as possible, should be selected.

VII

Control/Display Integration

User Dialogue

In the context of user-computer interaction, dialogue refers to the sequence of transactions which mediate user-system interaction [135]. Several types of dialogue design are available for developing an effective human-computer interaction. Many of these techniques are dependent on the task to be performed by the operator. That is, selection of an appropriate dialogue must require prior knowledge of the specific job requirements and the type of user. The following dialogues used in a variety of interactive applications are examined:

- Question and answer
- Form filling
- Menu design
- Command language
- Query language
- Natural language
- Expert systems

Question and Answer

The question and answer format is a computer-initiated sequence of transactions between the user and the system that provides explicit prompting in performing task and control activities for the unskilled, occasional user.

Guidelines	Research Support	Source
1. Question-and-answer dialogue should be used primarily for routine data entry tasks where the user has little or no training.	N	135
2. The data items should be known and their ordering constrained.	N	135
3. The computer response should be moderately fast.	N	135

Comment. Brief question-and-answer sequences can be used to supplement other dialogue types for special purposes, such as for log-on routines or for resolving ambiguous control or data entries.

Form Filling

This is a computer-initiated sequence of transactions between the user and the system that provides flexibility in data entry activities for moderately trained users.

Guidelines	Research Support	Source
1. Form-filling dialogue should be used when some flexibility in data entry is needed, such as the inclusion of optional as well as required items, where users will have moderate training, and/or where computer response may be slow.	L	135

Menu Design

This is a popular user-initiated sequence of transactions between the user and the system that provides explicit, single-choice task and control activities for the unskilled user.

Typically, the computer presents a series of options, and the user selects one or more from the options presented. They may be single-page or multi-page, titled, or layered [152]. Menus may also be hierachical and cyclic
Guidelines for option codes and applications for menus are as follows.

Guidelines	Research Support	Source
Menus--Option Codes/Applications		
1. Selection should be accomplished by keyed entry of corresponding codes or by other means such as programmed multifunction keys labeled in the display margin.	L	135
2. When menu selection is accomplished by code, that code should be keyed into a standard command entry area (window) in a fixed location on all displays.[a]	L	135
3. When control entries will be selected from a discrete set of options, those options should be displayed at the time of selection.	L	135
4. Displayed options should be worded in terms of recognized commands or command elements.[b]	L	135
5. If menu selections must be made by keyed codes, each code should be the initial letter (or letters) of the displayed option label rather than an arbitrary number.[c]	L	135
6. If letter codes are used, those codes should be used consistently in designating options at different steps in a transaction sequence.	L	135
7. Menus should be used to minimize training needs.	Y	44
8. Menus should be used when users have little or no typing skills.	Y	44
9. Menus should be used when the system has a limited keyboard.	Y	44

[a] In effect, the command entry area should be positioned to minimize user head/eye movement between the display and the keyboard [135].

[b] Where appropriate, sequences of menu selections should be displayed in an accumulator until the user signals entry of a completely composed command [135].

[c] An exception to Guideline 5, options might be numbered when a logical order or sequence is implied; and, when menu selection is from a long list, line number might be an acceptable alternative to letter codes [135].
Letters are easier than numbers for touch typists; options can be reordered on a menu without changing letter codes; it is easier to memorize meaningful names than numbers, and so letter codes can facilitate a potential transition from menu selection to command language when those two dialogue types are used together [135].
User System Interface (USI) designers should not create unnatural option labels, just to ensure that the initial letter of each will be different. There must be some natural difference among option names, and a two- or three-letter code can probably be devised to emphasize that difference [135].

	Guidelines	Research Support	Source
	Menus--Option Selection		
1.	Computer response should be fast.[a]	N	135
2.	Each menu display should require just one selection by the user.[b]	L	135
3.	Displayed menu options should be listed in a logical order; if no logical structure is apparent, then options should be displayed in order of their expected frequency of use, with the most frequent listed first.[c]	L	135, 44
4.	Displayed menu lists should be formatted to indicate the hierarchic structure of logically related groups of options, rather than as an undifferentiated string of alternatives.[d]	L	135
5.	If menu options are grouped in logical subunits, those groups should be displayed in order of their expected frequency of use.	L	135
6.	Use the same color for menus within the same group.	Y	44

[a] When display output is slow, as for a printing terminal or for an electronic display constrained by a low-bandwidth channel, it may be tiresome for a user to wait for display of menu options [135].

[b] Novice users will be confused by any more complicated procedure, such as a Chinese menu requiring one choice from Column A, two from Column B, etc [135].
The two actions should be compatible in their design implementation. If the cursor is positioned by keying, then an ENTER key should be used to signal control entry. If the cursor is positioned by light pen, then a dual-action trigger on the light pen should be provided for positioning and control entries [135].

[c] Logical grouping of menu options will help users learn system capabilities [135].

[d] If the first menu option is always the most likely choice, then for some applications it may be useful for efficiency of sequence control if a null entry defaults to the first option. If that is done, it should be done consistently [135].

Comment. Readers are advised to keep descriptive labels from being confused with option labels. Likewise, limiting the number of options on a menu to only those appropriate to the transaction will aid in limiting user errors. Hierarchical ordering and clustering of options are also approved methods for presentation of long menus. Parsimony is a good rule, try to limit the number of steps a user must take in making selections.

Guidelines for menu navigation are as follows:

Guidelines	**Research Support**	**Source**
Menu--Navigation		
1. When hierarchic menus are used, the user should be given some displayed indication of current position in the menu structure.[a,b]	L	135
2. When hierarchic menus are used, a single key action should permit the user to return to the next higher level.	L	135

3.	Menus provided in different displays should be designed so that option lists are consistent in terminology and ordering.	L	135
4.	Experienced users should be provided means to bypass a series of menu selections and make an equivalent command entry directly.	L	135
5.	When a user can anticipate menu selections before they are presented, means should be provided to enter several stacked selections at one time.	L	135
6.	Menu displays for a system still under development might indicate future options not yet implemented, but those options should be specially designated in some way.	L	135

[a] A single menu that extends for more than one page will hinder learning and use. The USI designer can usually devise some means of logical segmentation to permit several sequential selections among few alternatives instead of a single difficult selection among many [135].

[b] One possible approach would be to recapitulate prior (higher) menu selections on the display. If routine display of path information seems to clutter menu formats, then a map of the menu structure might be provided at user request as an optional HELP display [135].

Comment. A menu-driven dialogue is perhaps the most popular interactive technique available. In summary, menu-selection dialogue is most applicable if the users are inexperienced. It is also a benefit when the task requirements are routine and/or when the command set is so large that users are not likely to commit all command sets to memory.

Command Language

If designed properly, command language is a user-initiated sequence of transactions between the user and the system; that provides high flexibility and efficient performance in task and central activities for highly skilled users. Although common functions are supported by today's information retrieval systems, there is little uniformity in vocabulary and syntax between command languages. One recent effort to achieve standardization has been to develop the NISO standard Z39.58 [108].

The NISO Z39.58 standard, under development, recommends the following commands being considered in the design of commands structure for the user interface:

- EXPLAIN – To obtain nonsession specific information about the system, its use, and its data bases
- HELP – To directly obtain on-line assistance
- START – To initiate a session, reinitiate all default values and settings
- STOP – To terminate a session
- CHOOSE – To select files or data bases to be searched
- FIND – To invoke a search statement
- DISPLAY– To view on-line at user's terminal
- PRINT – To request off-line printing
- SORT – To arrange records in search results sets by specified fields
- SAVE – To save all previous statements in the session
- KEEP – To identify and save within a session

	Guidelines	Research Support	Source
1.	When command language is used for control entry, an appropriate entry area should be provided in a consistent location on every display, preferably at the bottom.[a]	L	135
2.	The words chosen for a command language should reflect the user's point of view and not the programmer's.	L	135
3.	Abbreviation of entered commands (i.e., entry of the first one to three letters) should be permitted to facilitate entry by experienced users.[b]	L	135

4.	Do not label two commands DISPLAY and VIEW when one permits editing displayed material and one does not.[c]	L	135
5.	The user should be able to request display of a file by name alone without having to enter any further information such as file location in computer storage.	L	135
6.	The user should be able to request prompts as necessary to determine required parameters in a command entry or to determine available options for an appropriate next command entry.[d]	L	135
7.	The user should be able to enter commands without punctuation.[e]	L	135, 44
8.	Neither the user nor the computer program should have to distinguish between single and multiple blanks in a command entry.	L	135
9.	The computer should be programmed to recognize common misspellings of commands and to display inferred correct commands for user confirmation rather than requiring reentry.[f]	L	135
10.	When a command entry is not recognized, the computer should initiate a clarification dialogue rather than rejecting the command outright.[g]	L	135
11.	The system shall accept user input without discriminating between upper and lower case.	L	44
12.	Command language assumes highly experienced and trained users.	L	108

---

[a] Adjacent to the command entry area there should be a defined display window used for prompting control entry, for recapitulation of command sequences (with scrolling to permit extended review), and for mediating question-and-answer dialogue sequences (i.e., prompts and responses to prompts) [135].

[b] Variable abbreviation, i.e., keying only enough characters of a command to uniquely identify it, should probably not be used when the command set is

changing. For the user, an abbreviation that works one day may not work the next [135].

[c] In general, do not give different commands semantically similar names, such as SUM and COUNT [135].

[d] In some applications it may be desirable to let an inexperienced user simply choose a general prompt mode of operation, where any command entry produces automatic prompting of (required or optional) parameters and/or succeeding entry options [135].

[e] If punctuation is needed, perhaps as a delimiter to distinguish optional parameters or the separate entries in a stacked command, one standard symbol should be used consistently for that purpose, preferably the same symbol (slash) used to separate a series of data entries [135].

[f] This practice should permit a sizable reduction in wasted keying without serious risk of misinterpretation. The necessary software logic is akin to that for recognizing command abbreviations [135].

[g] Poorly stated commands should not simply be rejected. Instead, the computer should be programmed to guide the user toward a proper formulation, preserving the faulty command for reference and modification, and not require the user to rekey the entire command just to change one part [135].

Comment. Unlike menu dialogue, command language dialogues should be reserved for sophisticated and highly trained users and where computer response is expected to be relatively fast.

Query Language

Query language is a user-initiated sequence of transactions between the user and the system that provides flexible and efficient performance in information retrieval tasks for moderately trained users.

Guidelines	**Research Support**	**Source**
1. Query language dialogue should be used as a specialized subcategory of general command language for tasks emphasizing unpredictable information retrieval (as in many analysis and planning tasks).[a]	L	135

2.	The organization of the query language should match the data structure perceived by users to be natural.[b]	L	135
3.	One single representation of the data organization should be established for use in query formulation, rather than multiple representations.[c]	L	135
4.	The need for quantificational terms in query formulation should be minimized or eliminated.[d]	L	135
5.	Use of operators subject to frequent semantic confusion, such as "or more" and "or less," should be minimized.	L	135

[a] All recommendations for command language design would apply equally to query languages [135].

[b] The users' natural perception of data organization can be discovered through experimentation or by survey [135].

[c] Beginning or infrequent users may be confused by different representational models.

[d] People have difficulty in using quantifiers unambiguously. When quantifiers must be used, it may be desirable to have the user select the desired quantifier from a set of sample statements so worded as to maximize their distinctiveness.

Comment. In conclusion, query languages should (a) display a restatement of the user's inquiry to assure correct interpretation of the user's intended meanings; (b) present user confirmation of system interpretation of meaning before the system executes the command; (c) display information in the form needed by the user even if the format differs from that contained in the data base or the form in which the data were originally entered; and (d) be such that the user's perception of the data base should be sufficiently structured so as to enable rapid identification of those parts in which the user is interested [99].

Natural Language

This is a developing user-initiated sequence of transactions between the user and the system that will provide flexibility and efficient performance in task and control activities for moderately skilled and, eventually, unskilled users.

Guidelines	**Research Support**	**Source**
1. Consider using some constrained form of Natural language in applications where task requirements are broad-ranging and poorly defined, where little user training can be provided, and where computer response will be fast.[a]	L	135

[a] For applications where task requirements are well-defined, other types of dialogue will prove more efficient [135].

Comment. Some form of restricted natural language is feasible when one cannot teach a command set and syntax or vocabulary size does not hinder problem formulation. Restricted natural language, though not a panacea, should be considered (a) when it is impossible to teach a formal query language to potential users and the system's task is narrow and well-defined or (b) when unsophisticated users must use a system with a moderate number of functions.

Natural language should be implemented into a system cautiously. Otherwise, ambiguities commonly employed in conversational English may tend to confuse the system, making it difficult if not impossible to operate.

Expert systems

Software programs that evaluate structured information and draw conclusions concerning a given subject usually fall under the category of expert systems [157]. Since it is difficult to define an expert system, it is perhaps more informative to contrast how expert system differ from conventional programs:

- Expert systems typically perform tasks previously performed by a knowledgeable human specialist;
- Expert systems are usually maintained by knowledge engineers, whereas conventional programs are maintained by programmers;
- Expert systems rely on heuristics for their structure, while conventional programs rely on algorithms [63].

In the chemical/nuclear process industry, expert systems have potential applications for event diagnosis, safety function assessment, procedure generation, site emergency actions, maintenance planning, and monitoring technical specification operating limits [160]. The benefits of an expert system can be realized through

- The capability for having full-time expert consultation,
- Improved operating efficiency,
- Reduction in operator workload via on-line computer aided capability for integration and analysis of multiple information sources, and
- Improved safety to equipment and personnel.

However, the benefits for expert systems must also be weighed against the following potential risks:

- Overreliance on the expert system and less reliance on human judgment,
- Discrepancy between the instructional aiding provided by the expert system and reality,
- Unidentified errors in the software,
- Unforeseen events (i.e., those events not modeled or incorporated into the expert system's knowledge base).

In spite of the risks, it is predicted that expert systems will play a prominent role in future process control systems. One aspect of expert systems that has been gaining new prominence in recent years involves the design of the user interface. Although this field is in its infancy, the user interface cannot be overlooked as an integral component of the expert system. Some recent issues and observations for designing a user interface for an expert system are shown in Table 12 [105].

The following basic human factors considerations should be taken into account in the development and implementation of user system interfaces for an expert system.

Guidelines	Research Support	Source
General		
1. The system should make it easy and natural for a user to inquire about any details desired.	L	63
Expert System Dialogue		
2. The system should support a flexible dialogue that permits either the user or the expert system to initiate an action or request for information without cancelling an ongoing transaction.	L	149
3. The user-expert system dialogue should be flexible in terms of the type and sequencing of user input it will accept.	L	149
User-Assistance		
4. The system should be capable of supporting speculative analysis (e.g., what-if scenarios) by providing a "reconnoiter mode" that allows the user to investigate the effects of an action without actually implementing the action.	L	149
Problem Definition		
5. The knowledge required to perform all functions allocated to the expert system should be directly accessible by the expert system. Requirements for the expert system to query the user to obtain information for routine functions should be minimized.	L	149
6. The capability for the user to supercede the current request for information from the expert system in order to input information related to a different transaction should be provided.	L	149

TABLE 12. GENERAL ISSUES FOR DEVELOPMENT OF EXPERT SYSTEMS - PRELIMINARY FINDINGS AND OBSERVATIONS. (Source: Adapted from Reference [105])

- Operators will not use a computer aid for tasks they can do themselves.
  - Aid should be aimed at a task upon which operators will accept help.
- A mismatch between computer capability and subject expectation can cause confusion at times and may reduce the usefulness of the aid.
  - Operator stereotypes of what a computer can do must be considered when designing an aid to the operators' thinking, i.e., an artificial intelligence application.
- Operators believe data provided by a computer.
  - Information displayed on a computer must be correct since it will be believed.
- An integrated CRT piping and instrumentation diagram mimic/touch panel can be effective for display and control.
  - Mimic/touch panel arrangements have a higher user acceptance.
- Operating priorities developed in aid design can be helpful if the aid is implemented or not.
  - Development of an operator aid provides clarification of the operators task, thus making it easier to perform.

Information Display

7. The expert system should have the capability to graphically represent its rules network. This capability should be available to the user as an adjunct to the explanation subsystem.[a,b] Y 149

Consultation

8. The expert system should automatically record all rules invoked during a consultation. Following a consultation, the explanation facility should be capable of recalling each invoked rule and associating it with a specific L 149

	event (i.e., question or conclusion) to explain the rationale for the event.		
9.	The expert system should be able to respond to user requests to clarify or restate questions and assertions.	L	149
10.	At any point during a transaction, the expert system should be able to explain which problem-solving strategy is being employed, why a particular strategy was selected, and the current status of the application.	L	149
	Level of Explanation		
11.	The level of detail of information presented as part of an explanation or justification should be under the control of the user. As a minimum, the user should be able to specify three levels of detail: rules only, brief explanations, and detailed explanations.	L	149

[a] To the extent possible, graphics should portray system status through the use of color, highlighting, or other coding technique [149].

[b] Reference [63] states, "Graphics representations may be particularly helpful. Likewise, displays that allow the user to follow the systems reasoning process may be a key to 'selling' the system to users."

Comment. The state of the art for user interface design features for expert systems is relatively new and continuously evolving. The availability of human factors guidelines for expert systems is limited and mostly of a general nature. Without a rigorous cookbook methodology in place, the need for thorough front-end analysis as the precursor to detailed design of the expert system cannot be overestimated. In the early phases of conceptual design, expert system development should be based on (a) user requirements, (b) preferred dialogue and knowledge of engineering requirements, (c) operational requirements; and (d) mental models employed by the human expert and user [149]. In addition, a detailed description of the functional transactions to be performed between user and expert system should be developed prior to the development of the internal structures. During the front-end analysis, particular attention should be placed on the suitability of the tasks to be performed by the expert system. Buchanan, Barstow, Bechtel, et. al. [19] refer to the following maxims regarding task suitability for construction of an expert system: (a) Focus on a narrow specialty area that does not involve a lot of common sense reasoning and knowledge; (b) Don't select a task that is too easy nor too difficult for human experts; (c) Be sure that the task (both inputs and outputs) is very clearly defined; and (d) Get long-term commitment from the subject matter expert(s) needed to define the tasks.

All too often, many design efforts for expert systems have been failures or were developed at excessive expense because the designers chose to take a bottom-up as opposed to a top-down development strategy. Expert systems that follow the bottom-up approach rarely get past the prototype stage. Usually the project runs out of funds, or a manager's patience, before the final product is ever implemented.

In addition to the need for thorough front-end analysis, the user interface designer of expert systems. should seriously consider the power of graphical display mediums for information presentation.

System Feedback

When interacting with a terminal, immediate feedback is essential for establishing the user's confidence, satisfaction, and ability to perform his or her specific tasks effectively. At all times, the user must know where he is, where he has been, and where he can go. Throughout this interaction process, the user must also know whether the system is operating and recognize when immediate feedback is delayed or interrupted.

The variables discussed in this section are

- Display update rate
- Response time
- System status indication
- Routine status information
- Performance/job aids

Display Update Rate

The term update rate is defined as the frequency of CRT update per unit time and the amount of time delay between the change of a system parameter and its graphic or numeric update on a CRT screen. For example, if the hot leg changes in the plant but the CRT display update lags that change by 5 s, the update rate is said to be 5 s. There are numerous factors that significantly influence the update rate of displayed parameters to operators while interacting with a computer system. Some of these variables include sampling rate, memory, I/O overhead, cable length, type of CRT terminal, computer architecture, complexity of averaging algorithm, number and type of sensing transducers, and port configuration [5].

Relevant guidelines are presented below.

Guidelines	Research Support	Source
1. Update rates for continuous, real-time tracking tasks should not exceed 0.5 s.[a]	Y	9
2. In general, update rates should not exceed 3 s.[b]	L	9

[a] Feedback delays of visual information greater than 0.5 s will seriously impede a continuous real-time control tracking task [137].

[b] In informal research conducted at the INEL Loss-of-Fluid Test (LOFT) facility using interactive graphic displays and LOFT operators, Banks found that update rates of 3 s were acceptable to the operators. These observations were made, however, under conditions where all operating personnel were expecting a transient to occur. Because of this expectancy effect and also because the control task was not randomly ordered, the observations are more directed at operator preference than performance [9].

Comment. The many constraints which influence the update rate of displayed parameters make it difficult if not impossible to establish minimum values. Obviously, the shorter the update time the better, but it may be realistic to identify update rates only for critical tasks, e.g., those tasks wherein a delay of information will seriously impede safe operation of a physical control process [9]. Therefore, the update rate should be analytically determined by systematically examining (a) the type of process being controlled, (b) the type of tasks performed by operators, (c) safety critical value of control parameter, (d) consequences of delayed information on a worst-case basis, and (e) temporal response requirement of a system.

If such analysis is not feasible, a good rule of thumb might be the observational derived criteria for update rates collected at INEL (update rates of 3 s were acceptable to the operators). It should be noted that these data must be interpreted and implemented with caution due to the lack of formal experimental validation. The work performed by Smith and Smith [137] is the product of a more formalized study, but the specific experimental task and findings may not be directly generalizable to process control applications.

Response Time

The concept of computer response time most often refers to the time that the user must wait for a computer response following a command. In other words, it is the elapsed time between a user request and a meaningful reply [5]. The ideal situation is, of course, when there is no response time at all. Since this is not always feasible, some acceptable guidelines should be established based on tradeoff analysis between the user's ability to perform his or her task and hardware/software limitation. The determination of acceptable response times are highly task-specific. That is, the user's expectations may differ between situations. As a result, much of the literature cites a set of response times as they apply to specific tasks. The operator's willingness to wait is functionally related to the perceived complexity of the task and the time when the request was made. Other research points out that it may not only be the magnitude of the response time but the variability of the delays which significantly impact operator performance and acceptance of a system.

Guidelines	Research Support	Source
1. Response times should be within the maximums shown in Table 13.	L	47
2. Response time deviations should be less than one-half the mean response time.[a]	Y	54

[a] Carbonell et al. [24] point out that it is the variability of delays, not their magnitude, which is frequently the most distressing factor to users. From a consistency standpoint, a good rule of thumb is that the response time deviations should never be more than half the mean response time (e.g., if the mean response time is 4 s, the variation should be confined to the range of 3 to 5 s; a 2 s deviation).

Comment. Adequate values for determining response times are highly task-specific, and reliable research data are limited. The values cited in Engel and Granda [47] (shown in Table 13) are only educated guesses or armchair estimates. However, for lack of more concrete data, those values can also be construed as reasonable if the specific task requirements are known. If specific tasks are not identified, the most often recommended armchair estimates range from 2 to 4 s maximum system response time [94].

System Status Indication

This term is used to describe a method, technique, or signaling device (auditory or visual) that permits the operator of a CRT terminal or visual display to easily, accurately, and rapidly detect whether the computer is functioning normally.

Guidelines	Research Support	Source
1. An indication that the computer or control panel is functioning normally should be provided on the CRT display.[a]	N	7, 85

TABLE 13. ACCEPTABLE RESPONSE TIMES (Source: Reference [126], as presented in Reference [47])

	User Activity	Maximum Response Time (s)
1.	Control activation (for example, keyboard entry)	0.1
2.	System activation (system initialization)	3.0
3.	Request for given service:	
	Simple	2
	Complex	5
	Loading and restart	15-60
4.	Error feedback (following completion of input)	2-4
5.	Response to ID	2
6.	Information on next procedure	5
7.	Response to simple inquiry from list	2
8.	Response to simple status inquiry	2
9.	Response to complex inquiry in table form	2-4
10.	Request for next page	0.5-1
11.	Response to "execute problem"	15
12.	Light pen entries	1.0
13.	Drawings with light pens	0.1
14.	Response to complex inquiry in graphic form	2-10
15.	Response to dynamic modeling	--
16.	Response to graphic manipulation	2
17.	Response to user intervention in automatic process	4

[a] Reference [85] recommends continuous display of a sweeping vector (similar to the second hand of a watch) in the upper quadrant of the screen. Reference [7] recommends a 0.5-in.-diameter circle or 0.5-in. square that pulses from black to white at a rate of 1 pulse/s.

Comment. The criterion above describes passive alarms which may not have the same attention-getting capability as an active failure alert system incorporated in the CRT display. For example, the installation of a self-powered circuit that would emit a periodic beeping tone when the system fails would have more attention-getting characteristics than a passive visual stimulus. Therefore, the authors recommend that an active alert system take precedence over a passive mode when feasible. A digital clock placed in a corner of a screen is not an adequate mode for alerting the operators of a system failure. There is observational evidence that indicates the clock, as a stand-alone status indicator, does not provide adequate attention-getting quality and should not substitute for the criterion mentioned above.

Routine Status Information

The various mechanisms for informing the operator of current system status are examined in this section. All too often, the operator must know the state of displayed information throughout the interactive process.

It should be noted that many of these techniques encompass a substantial number of other variables addressed in this document. Therefore, the reader is encouraged to review supporting guidelines information. The sections on error statements and messages are recommended.

Guidelines	Research Support	Source
1. When system functioning requires the operator to stand by, periodic feedback should be	L	150, 5 135

provided to the operator to indicate normal system operation and the reason for the delay.[a]

2.	When a process or sequence is completed by the system, positive indication should be presented to the operator concerning the outcome of the process and requirements for subsequent operator actions.	L	150
3.	If at any time the keyboard is locked or the terminal is otherwise disabled, that condition should be signaled by disappearance of the cursor from the display and (especially if infrequent) by some more specific indicator such as an auditory signal.	L	135
4.	Status information should be available indicating current load (multiple users assumed) and/or current system performance.[b]	L	135
5.	Relevant status information for external systems should be available to the user.	L	135
6.	When time tagging information is important, date-time signals should be available to users as an annotation on displays.[c]	L	135
7.	Status information should be available concerning the current status of alarm settings, in terms of dimensions/variables covered and values/categories established as critical.[d]	L	135
8.	Every user input should consistently produce some perceptible response output from the computer.[e]	L	135
9.	Computer response to user entries should be rapid, with consistent timing as appropriate to different types of transactions.	L	135
10.	Following user interrupt of data processing, an advisory message should be displayed assuring the user that the system has returned to its previous status.	L	135

---

[a] Reference [135] states, "After making an entry to the computer, the user needs feedback to know whether that entry is being processed properly. Delays in computer response longer than a few seconds can be disturbing to the user, especially for a transaction that is usually processed immediately. In such a case, some intermediate feedback should be provided, perhaps an advisory message that processing has been initiated, and ideally with an estimate of how long it will take to complete."
Reference [5] states, "Keep in mind that providing a timely response to users may mean having the computer present a status message. Users sitting at the terminal and waiting for a response may wonder if the computer is still working, if the terminal is still connected to the computer, if the computer lost the input, or if the computer is in a never-ending loop. A short message–STILL PROCESSING--that appears every 10 s indicating that the computer is still working on the problem provides the user with some assurance."

[b] Such load information is primarily helpful when system use is optional, i.e., when a user can choose to defer work until low-load periods. But load status information may help in any case by establishing realistic user expectations for system performance [135].

[c] Date-time status might be displayed continuously or periodically, as on displays that are automatically updated, or by user request, depending on the application [135]. In some applications, request for date-time display can provide an innocuous means for a user to check on general system response.

[d] Alarm status information will be particularly helpful in monitoring situations where responsibility may be shifted from one user to another [135].

[e] Keyed entries should appear immediately on the display. function key activation or command entries should be acknowledged either by evident performance of the requested action or by an advisory message indicating an action in process or accomplished. Unrecognized inputs should be acknowledged by an error message [135].

Comment. All of the above guidelines echo a similar theme: The operator must be provided information at all times giving an up-to-date account of the systems. status, i.e., what is the system doing? A variety of techniques is available for satisfying the criteria.

Performance/Job Aids

Bailey [5] provides the most precise definition of a performance or job aid: "Performance aids are devices that store information for immediate use." They can be in either written hardcopy form or computer based. This section is devoted to those performance aids adapted for a CRT-generated medium. Performance aids are similar to step-by-step proceduralized instructions. A major difference, however, resides in the level of detailed information. Performance aids are usually characterized by having less to read in a quick-access format. Performance aids should be considered as a supplement to detailed procedures and/or training. Only those items not redundant with previous sections on menu display and user dialogue are presented below.

Guidelines	Research Support	Source
1. Specific user guidance information should be available for display at any point in a transaction sequence.[a]	L	135
2. To serve as a home base or consistent starting point at the beginning of a transaction sequence, a general menu of control options should always be available for user selection.	L	135
3. Hierarchic menus should be organized and labeled to guide the user within the hierarchic structure.[b]	L	135
4. Control options that are generally available at any step in a transaction sequence should be treated as implicit options, i.e., need not be included in a display of step-specific options.[c]	L	135
5. The computer should be programmed to provide prompting, i.e., to display advisory messages to guide users in entering required data and/or command parameters.[d]	L	135
6. When users vary in experience (which is often the case), prompting should be an optional guidance feature that can be selected by novice users but can be omitted by experienced users.[e]	L	7

7.	When the results of a user entry are contingent upon context established by previous entries, some indication of that context should be displayed to the user.	L	135
8.	Implicit cues for data entry should be provided by consistent and distinctive formatting of data fields.[f]	L	135
9.	Following computer generation of display output, the cursor should automatically be positioned on the display in a location consistent with the type of transaction.[g]	L	135
10.	Reference material should be available for on-line display to the user describing system capabilities and procedures.[h]	L	135
11.	In applications where a user may employ command entry, the computer should provide an on-line command index to help guide user selection and composition of commands.[i]	L	135
12.	A complete dictionary of abbreviations used for data entry, data display, and command entry should be available for on-line user reference and in system documentation.[j]	L	135
13.	When codes are assigned special meaning in a display, a definition should be provided at the bottom of the display.[k]	L	135
14.	In system applications where it is warranted, the user should be able to request a displayed record of past transactions in order to review prior actions.	L	135
15.	In addition to explicit aids (labels, advisory messages) and implicit aids (cueing) provided in user interface design, there should also be a capability for a user to request further on-line guidance by a request for HELP.[l]	L	135

16. When an initial HELP display provides only summary information, more detailed explanations should be available in response to repeated user requests for HELP.[m]	L	135
17. Novice users should be able to browse on-line HELP displays, just like a printed manual, to gain familiarity with system functions and operating procedures.	L	135
18. For many system applications, an on-line training capability should be provided to introduce new users to system capabilities and to permit simulated hands-on experience in data handling tasks.	L	135

---

[a] Do not require a user to remember information currently displayed. The user should not have to remember what actions are available or what action to take next [135].

[b] Users will learn menus more quickly if a map of the menu structure is provided as HELP [135].

[c] The user may be expected to remember continuously available options, once they have been learned, without their specific inclusion in a display of guidance information. Perhaps the best design expedient is to implement implicit options on appropriately labeled function keys, which will aid user learning and provide a continuing reminder of their availability [135].

[d] Prompting in advance of data/command entry will help reduce errors, particularly for inexperienced users. If a default value has been defined for null entry, that value should be included in the prompting information [135].

[e] Flexibility in prompting can also be provided by multilevel HELP options, so that additional guidance information can be obtained if the standard prompt is not adequate [135].

[f] Consistent use of implicit prompting cues can sometimes provide sufficient guidance to eliminate the need for more explicit advisory messages [135].

[g] Consistent cursor positioning will provide an implicit cue for user guidance [135].

[h] On-line access to a description of system structure, components, and options will aid user understanding. On-line guidance can supplement or, in some instances, substitute for off-line training An investment in designing

user aids may be repaid by reduced costs of formal training as well as by improved operational performance [135].

[i] Such a command index may help the user to phrase a particular command, but will be more generally useful as a reference for discovering related commands and learning the overall command language [135].

[j] In applications where users can create their own abbreviations, as in the naming of command macros, it will be helpful to provide aids for users to create their own individual on-line dictionaries [135].

[k] This practice will aid user assimilation of information, especially for display codes that are not already familiar [135].

[l] It is difficult for an interface designer to anticipate the degree of prompting that may be required to guide all users. Moreover, even when prompting needs are known, it may be difficult to fit all needed guidance information on a working display. A supplementary HELP display can be provided to deal with such situations [135].

[m] Designing the HELP function to provide different levels of increasing detail permits users to exercise some judgment themselves as to just how much information they want [135].

Comments. Despite the relative importance of job aids as an integral part of user dialogue, three problems often constrain their full capabilities:

1. They are not considered at all.
2. Performance aids are installed, but they are either incomplete or so difficult to access and use in the dialogue that the user fails to utilize them.
3. The original source material used to prepare the job aid is often of inferior quality. An on-line adaptation of a poorly prepared technical manual will not improve the quality of the information.

Specific design criteria for development of a performance aid from the top down is beyond the scope of these guidelines. However, implementation of the above guidelines should enhance the user-friendly capabilities of existing performance-aiding devices.

Software Security

Basically, the issue of software security encompasses two areas: (a) the need to protect data from unauthorized access and tampering and (b) the need to protect data from authorized users who may induce errors causing loss and trashing of data files. Unfortunately, there has also arisen a need to protect users from viruses. Viruses have appeared, as a software problem, with the advent of public bulletins where free ware and share ware is exchanged [143]].

Techniques and methods for ensuring that data can be protected from both external threat and user error are examined in the section on Data Protection/Data Security.

Most recently, Woods et al. [159] have identified 13 categories related to computer software and hardware security. These include personnel policies, system development, processing, and physical access parameters. Within each of these categories are a number of checklist items weighted by an index of applicability. This checklist approach holds promise as recent resources on computer security are rather limited.

Data Protection/Data Security

Computer systems used for process control often store data of a proprietary nature. As a result, the system designer is often tasked to develop a system that is secure from unauthorized tampering. Provisions must be implemented which will minimize the probability of damaging or losing data through data entry errors.

Mair et al. [86] have argued that computer controls/security are needed to reduce the impact of computer fraud on our local and national resources. Practitioners are cautioned that security can be breached at the level of the programmer as well as at the level of the data entry clerk or computer system user. Mair et al. go on to state that, "relatively few companies have sufficient internal controls to reliably prevent or detect acts of computer fraud or embezzlement."

Furthermore, security can reduce the destruction of assets and help prevent suspension of plant operations due to breaches in security. In some instances loss of computer capability may be grounds for statutory sanctions (fines brought on by judicial authorities).

	Guidelines	Research Support	Source
	Data Security		
1.	Data Security should be protected by automatic measures whenever possible, rather than by administrative procedures.	L	135
2.	User interface design should provide consistent procedures for data transactions, including data entry and error correction, data change, and deletion.	L	135
3.	Inputs to the computer, including data entries and control entries, should require explicit user actions.[a]	L	135
4.	When the result of user action is contingent upon prior selection among differently defined operational modes, mode selection should be continuously indicated to the user, particularly when user inputs in that mode might result in unintended data loss.[b]	L	135
5.	User interface design should deal appropriately with all possible control entries, correct and incorrect, without introducing unwanted data change.[c]	L	135
6.	For both data entry and control entry, the user should be able to edit composed material before initial entry and also before any required reentry.[d]	L	135
7.	For both data entry and control entry, the user should be required to resolve any detected ambiguity requiring computer interpretation.	L	135

8.	The user should be warned of potential threats to data security by appropriate messages and/or alarm signals.	L	135
	Policy-Related Issues		
9.	Computer security procedures should be understood by all staff.	L	159
10.	Computer security policies should be strongly supported by management.	L	159
11.	Design system documentation explicitly delegates controls to be used.	L	159
12.	Policy is established whereby employees do not discuss security procedures outside of the job environment.	L	159
13.	Unbeknownst to the user, the computer automatically logs user ID and keeps record of file access and work performed.	L	159
14.	The system is kept free of "Shareware" and other programs which may contain viruses.	L	Authors
	Physical Access		
15.	Personnel are conspicuous by virtue of the fact that they are required to wear a badge.	L	159
16.	Visitors are required to wear identification.	L	159
17.	Passwords are employed by all users (see Figure 39).	L	159
18.	Passwords are changed every 90 days.	L	159
19.	Passwords are changed every two weeks (high security access).	L	159
20.	Physical key locks are provided.	L	159

Figure 39. The use of an employee's individual keycard for giving access to a classified computer system.

Control Implementation

21.	Internal and external security audits are conducted on a regular basis.	L	86
22.	Commensurate with review of security, reliability of the system as a whole should be calculated.	L	86

23.	Data conversion procedures are subject to scrutiny.	L	86
24.	There are vertical controls, e.g., those between levels of the organization.	L	86
25.	There are horizontal controls, those between departments or agencies.	L	86
26.	Standards are in place which call for the use of controls.	L	86
27.	Construction practices should promote a fireproof and waterproof environment.	L	Authors

---

[a] Interface designers are sometimes tempted to contrive smart shortcuts in which one user action may automatically produce several other associated data changes, perhaps saving the user a few keystrokes. Since such shortcuts cannot generally be made standard procedures, they will tend to confuse novice users and so may be a potential threat to data protection [135]

[b] A use cannot be relied upon to remember prior actions. Any action whose results are contingent upon previous actions represents a potential threat to data protection [135].

[c] The user interface must be bullet-proofed so that an unacceptable entry at any point will produce no more significant computer response than an error message [135].

[d] This capability will permit a user to correct many entry errors before computer processing. When errors are made, the user will be able to fix them without having to regenerate correct items and risk introducing further errors [135].

Comment. The above guidelines should be a functional part of the overall security philosophy for process control systems. Further, it appears that many potential problems concerning data protection could be minimized if a top-level user-friendly system is a major consideration in the overall design.

VIII

WORKPLACE LAYOUT

Anthropometrics

Anthropometrics refers to human dimensions or measurements and takes into account height, reach, and girth characteristics. Paris and McConville [113] have stated that knowledge of this variability in body size is a fundamental aspect in the design of man/machine systems. In that same study, it was noted that many of the early nuclear power plant designs violated principles of access. Traditionally, studies have differentiated between male and female population characteristics. More recently, Gertman, Klinestiver and McConville [55] reviewed expected changes in anthropometric characteristics for the Naval aviator population of the future. In that same review; they examined the impact of disproportionality on functional reach and vision exterior to aircraft. Although not an exact parallel, disproportionality and population changes may impact reach envelope and lift absolutes for maintenance workers in process control as well. Much of the anthropometric research conducted to date was sponsored by the Air Force Medical Research Laboratory (AFMRL). In addition, specifications for systems exist in the form of tabled data found in MIL-STD-1472C, "Human Engineering Design Criteria for Military Systems, Equipmental and Facilities." [146]

In this section the following anthropometric variables are examined:

- Keyboard base
- Working level
- Keyboard home row
- Screen
- Viewing distance

- Footrest
- Reach Envelope
- Position and movement of the head
- Leg, knee, and foot room
- Screen orientation
- Chair
- Hardcopy printer
- Health and safety

There are essentially three basic types of workplaces where VDUs could be used in process control settings. Those are the sitting workplace, the standing workplace, and the sitting/standing workplace (see Figures 40, 41, and 42, respectively.) The decision as to which one is appropriate depends on the general control room layout, shift length, whether the job is action-oriented, or requires passive monitoring and encompasses a number of considerations.

Sitting workplaces are best in the following situations:

- All items needed in the short-term task cycle can be easily supplied and handled within the seated work space.
- The items being handled do not require the hands to work at an average level of more than 15 cm (6 in.) above the work surface.
- No large forces are required, such as handling weights greater than 4.5 kg (10 lb). (Large forces may be eliminated by using mechanical assists.)
- Fine assembly or writing tasks are done for a majority of the shift.

Standing workplaces will be the best alternative in the following circumstances:

- If the workplace or workstation does not have knee clearance for a seated operation.
- Objects weighing more than 4.5 kg (10 lb) are handled.
- High, low, or extended reaches, such as those in front of the body, are required frequently.
- Operations are physically separated and require frequent movement between workstations.

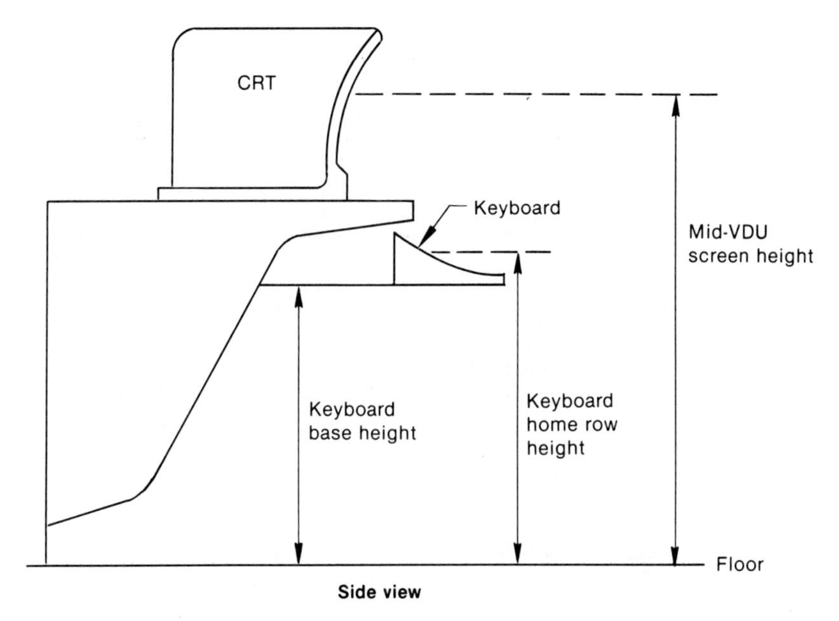

Figure 40. Seated VDU workplace (adapted from Reference [42]).

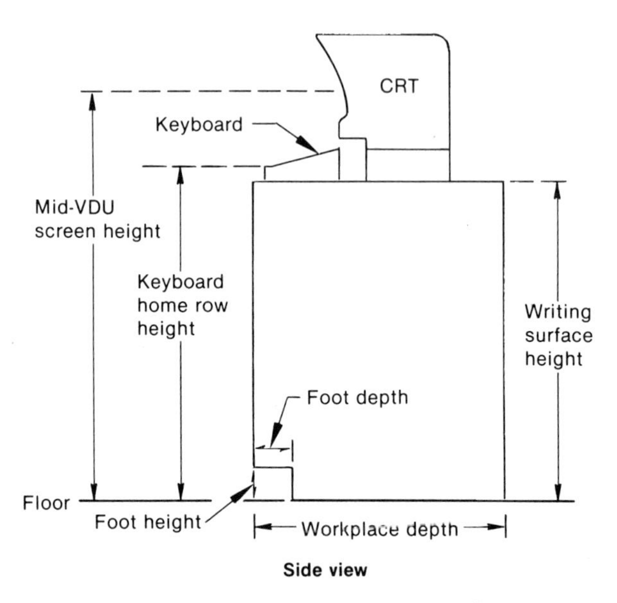

Figure 41. Standing VDU workplaces (adapted from Reference [42]).

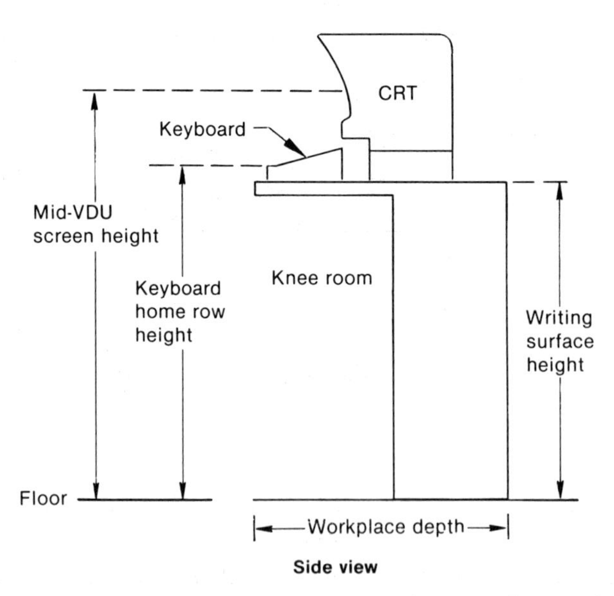

Figure 42. Sitting/standing VDU workplaces (adapted from Reference [42]).

- Downward forces must be exerted, as in wrapping and packing operations.

Sitting/standing workplaces should be considered in these instances:

- Repetitive operations are done with frequent reaches more than 41 cm (16 in.) forward and/or more than 15 cm (6 in.) above the work surface. operations would be done at a sitting workplace if it were not for the reach requirements.
- Multiple tasks are performed, some best done sitting and others best done standing. Provision for both may not be feasible owing to space constraints.

Because people are individuals and because they need to change positions periodically, adjustability within the recommended range is highly recommended. This adjustability accommodates those human traits. An example of an adjustable workstation is shown in Figure 43.

All these individual items, even if within guidelines, cannot guarantee a satisfactory workplace layout. The task to be performed, the manuals which are needed, and numerous other incidental considerations need to be addressed. This check probably goes beyond the VDU workplace to consideration of the overall control room or station design. How the VDU workplace fits into the overall system needs to be kept in mind. For a more detailed treatment of issues related to workstation design, the ANSI standard for Human Factors Engineering of VDTs is recommended [1].

Keyboard Base

Keyboard base refers to the work surface height upon which keyboards and other CRT related controllers are placed (see Figure 44).

Guidelines	**Research Support**	**Source**
1. The keyboard base height for a seated workplace should be from 56 to 77 cm (22 to 30 in.).[a]	Y	150, 5, 18, 119, 71

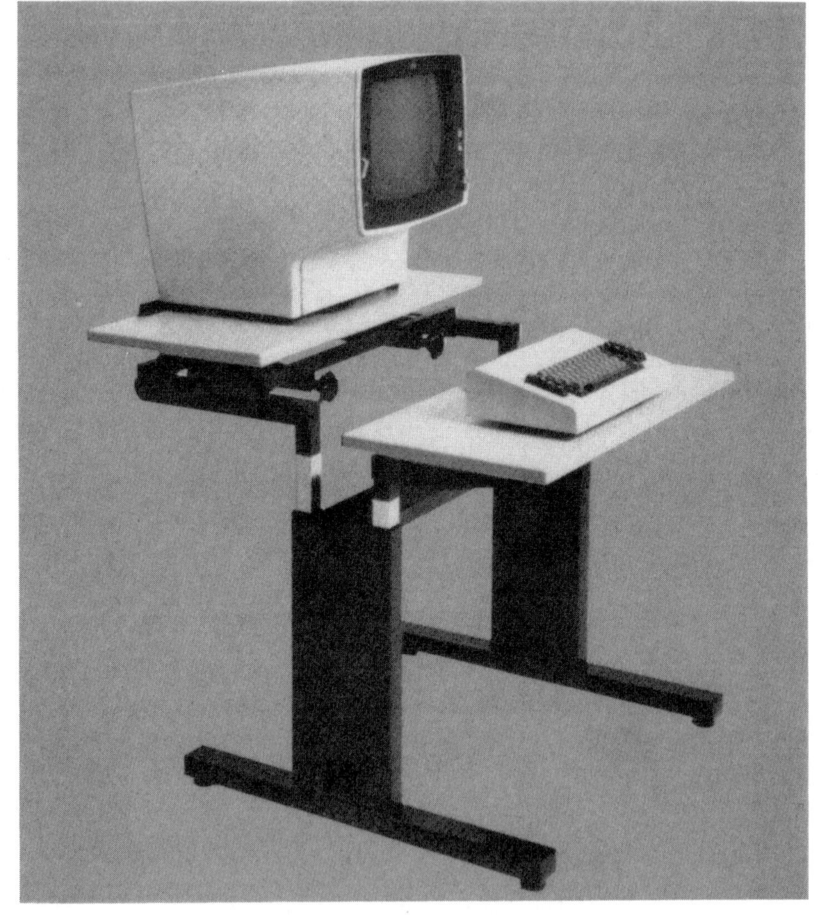

Figure 43. Adjustable VDU workstation. Capability is provided for independent height adjustment of the keyboard and screen surface.

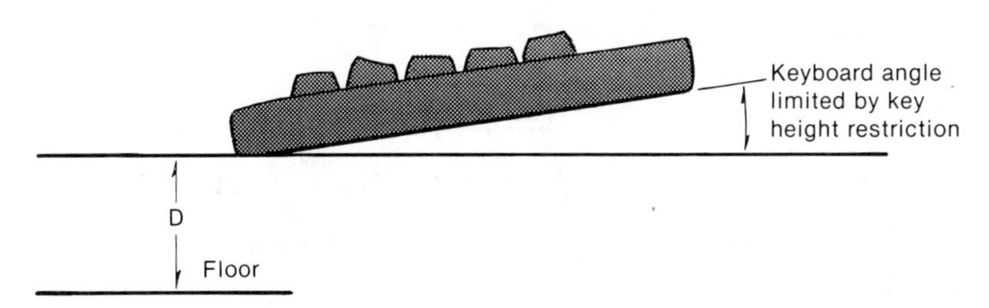

Figure 44. Keyboard base height (D) (adapted from Reference [111]).

2.	The keyboard base height for a standing or a sitting/standing workplace should be from 90 to 93 cm (35.5 to 36.5 in.).	Y	96

[a] Recommendations are based on short female (5th percentile) and tall male (99th percentile). Reference [18] recommends a range of 46 to 93 cm (18 to 36.5 in.).

Comment. The references cited provide the range of recommendations for keyboard base height assuming various workplace situations and population profiles. Again, adjustability over the recommended range to accommodate variations in personnel is strongly suggested.

Working Level

The working level is the working surface near a CRT which is utilized for writing, logging, or manipulative tasks not involving a CRT controller or keyboard (see Figure 45.)

	Guidelines	Research Support	Source
1.	Working level height for a sitting workplace should be from 66 to 81 cm (26 to 32 in.).	Y	150, 146, 23, 119,71, 49, 133

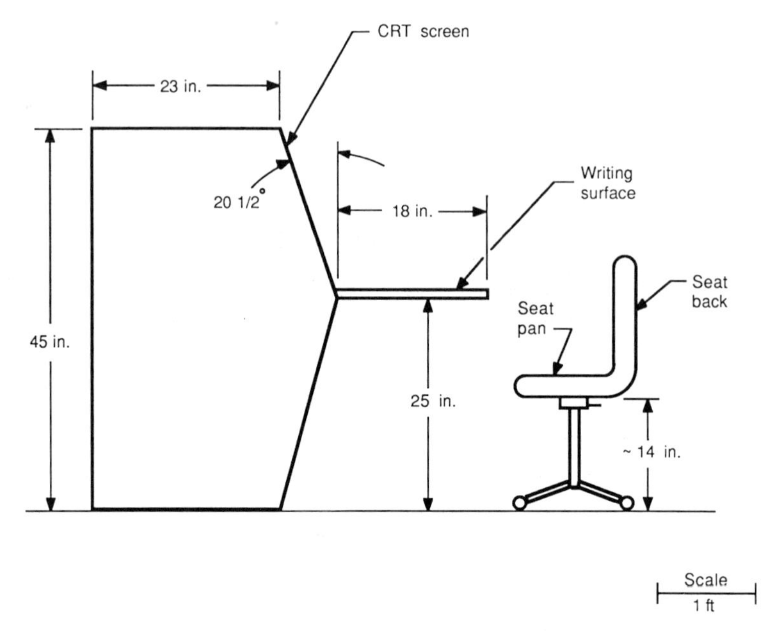

Figure 45. Side view of VDU workstation showing work areas.

2.	Working level height for a standing workplace should be from 90 to 107 cm (35.5 to 42 in.).	Y	146, 49, 135
3.	Working level height for a sitting/standing workplace should be from 90 to 102 cm (35.5 to 40 in.).	Y	146, 49, 135
4.	Working level width should be from 61 to 76.5 cm (24.4 in to 30.6 in), 76.5 cm preferred.	Y	150, 5, 146
5.	Working level depth should be from 41 to 64 cm (16.4 in. to 25.6 in.), 64 cm preferred.	Y	150, 5, 146, 46

Comment. These values reflect standard design practices in use industry-wide. Adjustability in the working surface is recommended.

Keyboard Home Row

The keyboard home row height is the distance from the floor (for seated operator) to the center of the home position for the right index fingertip on a Sholes or QWERTY keyboard (the center of the J key top) (see Figure 46).

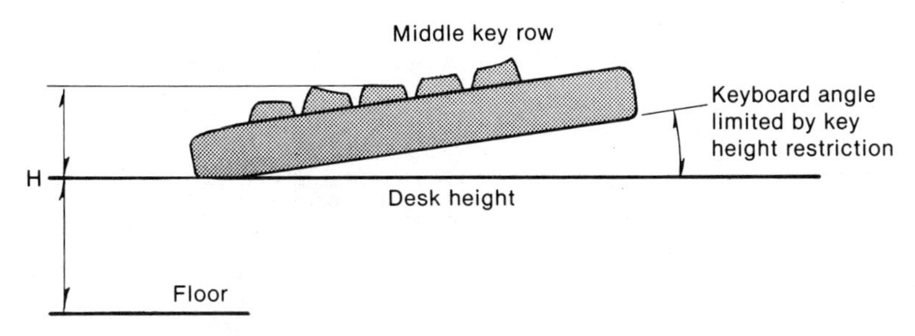

Figure 46. Keyboard home row height (H).

Guidelines	Research Support	Source
1. Keyboard home row height should be 66 to 78 cm (26 to 30.5 in.).	Y	23, 130

Comment. These values reflect standard design practices in use industry-wide. Adjustability to suit the individual, here as before, was recommended. Naturally, nonstandard seat heights will impact greatly the sanctity of this guideline.

Screen

The reference point for measuring screen height (display height) and viewing angle is that on the outermost surface of the screen midway between vertical and horizontal extremes of the portion used for displays. The display is the surface containing the images on which the eye is to focus and excludes filters, implosion covers, and similar added surfaces (see Figure 47).

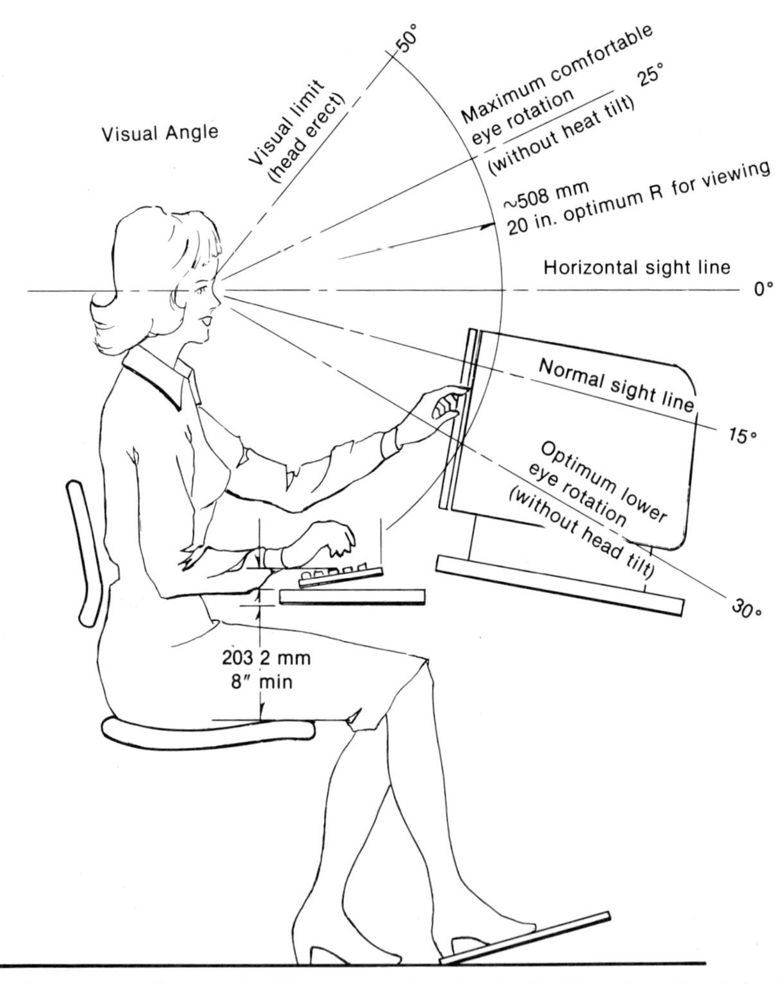

Figure 47. Display height (H) and visual angles for VDU workstation (adapted from Reference [23]).

Guidelines	Research Support	Source
1. The screen height for a seated workplace should be from 15 to 117 cm (6 to 46 in.), with 99 cm (39 in.) preferred.	Y	146, 17
2. The screen height for a standing workplace should be from 104 to 178 cm (41 to 70 in.).	Y	146
3. The screen viewing angle should be within 35 degrees of the horizontal line of sight, with about 15 degrees below the horizontal line of sight preferred.[a]	Y	150, 23, 18, 46, 74

[a] Reference [150] recommends that the viewing angle be within the upper limit of the visual field (75° above the horizontal line of sight) of the 5th percentile female, and the angle from the line of sight to the face plane is 45° or greater. The maximum lateral spread of controls and displays at a single-operator workstation should not exceed 72 in.

Comment. These values reflect standard design practices in use industry-wide. Adjustability to suit the individual, here as before, is recommended.

Viewing Distance

Viewing distance is the distance from the operator's eye to the surface containing the image being observed (see Figure 48).

Guidelines	Research Support	Source
1. The viewing distance should be 33 to 80 cm (13 to 30 in.), with 46 to 61 cm (18 to 24 in.) preferred.	Y	150, 42, 23, 18, 119, 71, 49, 74

Comment. All references fell within the range shown, with most showing a preference for the 18 to 24 in. range.

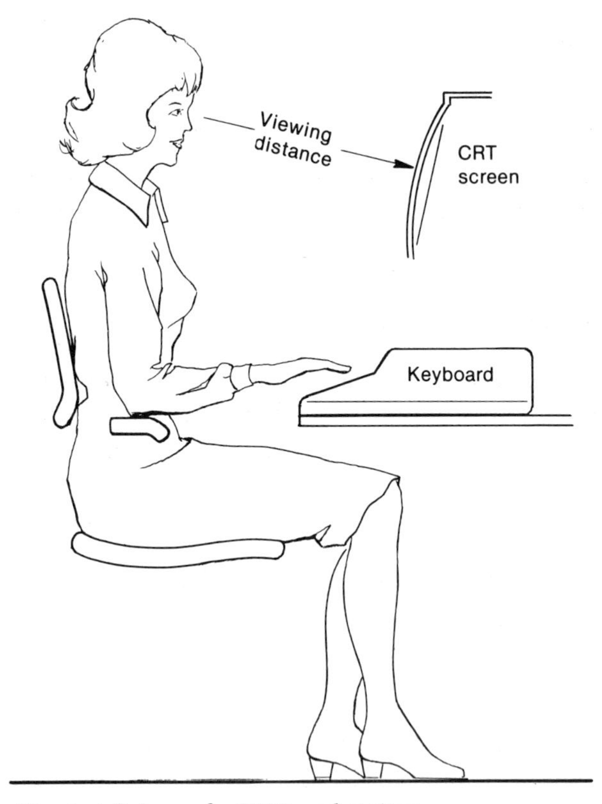

Figure 48. Viewing distance for VDU workstation.

Footrest

A footrest is a surface for the operator's use which adds leg comfort and reduces fatigue.

Guidelines		Research Support	Source
1.	The footrest should be 18 in. below the level of the seat and should be adjustable in 2-in. increments of height.	Y	150, 23, 46
2.	Rectangular footrests should be 30 cm (12 in.) deep by 41 cm (16 in.) wide.	Y	42
3.	Circular footrests should have a diameter of 18 in.	Y	150, 23, 46
4.	The footrest should be circular if it is part of the chair.	Y	150, 23, 46

Comment. The bottom line for footrests is, provide them if needed. Apparently, only shorter people express a desire for them.

Reach Envelope

Functional Reach Envelope (often referred to as thumb-forefinger grasping reach) is the measured distance from a constant reference point to the maximum reach limits of a stationary operator (see Figure 49).

Guidelines		Research Support	Source
1.	The functional reach envelope should be from 64 to 88 cm (25.2 to 34.6 in.).	Y	146, 113

Comment. The cited references fall within the above guideline for reach envelope.

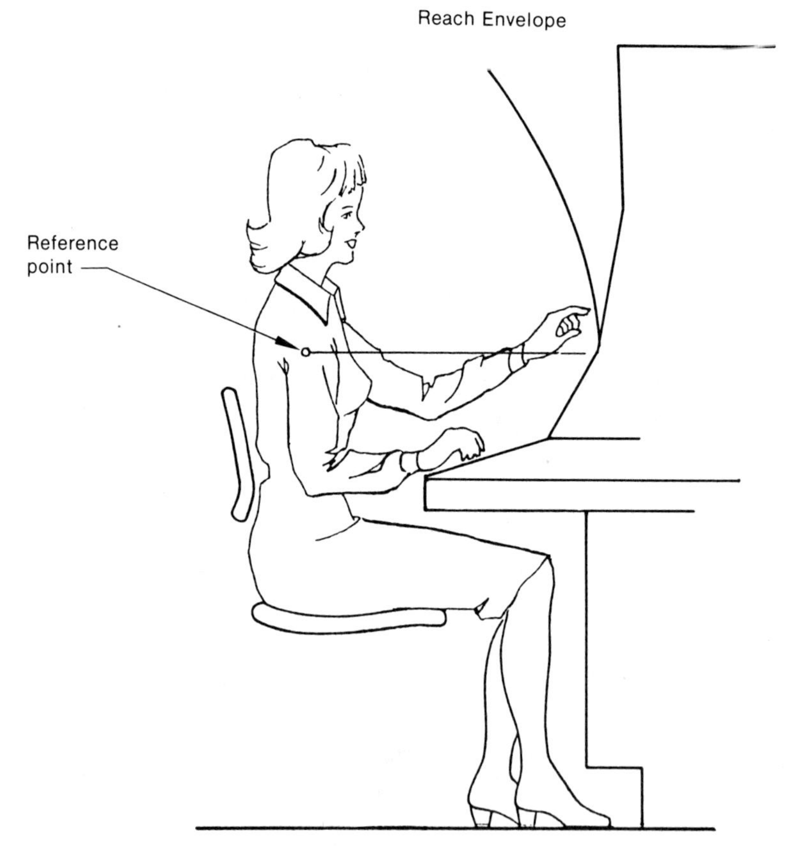

Figure 49. Functional reach envelope for seated operator.

Position and Movement of the Head

This pertains to the inclination of the head with respect to the horizontal plane.

Guidelines	Research Support	Source
1. The normal inclination angle of the head should be from 16 to 22 degrees.[a]	Y	23
2. A document holder should be provided to reduce head movement while keying data from a document.	Y	23

[a] The most comfortable viewing angle is between 32° and 44° below the horizontal plane. Head inclination is approximately one-half of this (16° to 22°) [23].

Comment. This guideline was developed with extensive keying in mind. However, the criteria are desirable (not mandatory) even where only occasional use of documents and keyboard/controls is employed.

Leg, Knee, and Foot Room

Relevant dimensional requirements for leg, knee, and foot room for the seated/standing operator are presented in this section.

Guidelines	Research Support	Source
1. For a sitting workplace, the following clearances should be provided:	Y	42, 46
a. knee clearance (depth) - 46 to 51 cm (18 to 20 in.)		
b. leg clearance (depth) - 100 cm (39 in.)		
c. leg clearance (width) - 51 cm (20 in.)		

2. For a standing workplace, the following clearances should be provided:	Y	42, 46
a. knee clearance (depth) - 10 to 50 cm (4 to 20 in.)		
b. foot clearance (depth) - 10 to 13 cm (4 to 5 in.)		
c. foot clearance (height) - 10 to 12 cm (4 to 5 in.)		
d. foot clearance (width) - 51 to 70 cm (20 to 28 in.)		

Screen Orientation

Screen Orientation is the angle the screen has with respect to straight-ahead horizontal viewing (see Figure 50).

Guidelines	Research Support	Source
1. Screen orientation should be no greater than 45 degrees away from or toward the operator, with 15 degrees away from the operator preferred.[a]	Y	42, 45, 18, 96, 17
2. Screen orientation should be adjustable.	Y	18, 96, 17

[a] The tilt angle of a display should consider the trade-off between reduction in the angular size of the symbols and glare [74].

Comment. An adjustable CRT angle is desirable because it allows compensation for varying glare conditions (as the control room is built and modified). If it is certain that glare can be adequately controlled (through use of dimmers, etc.), a fixed-angle CRT is acceptable.

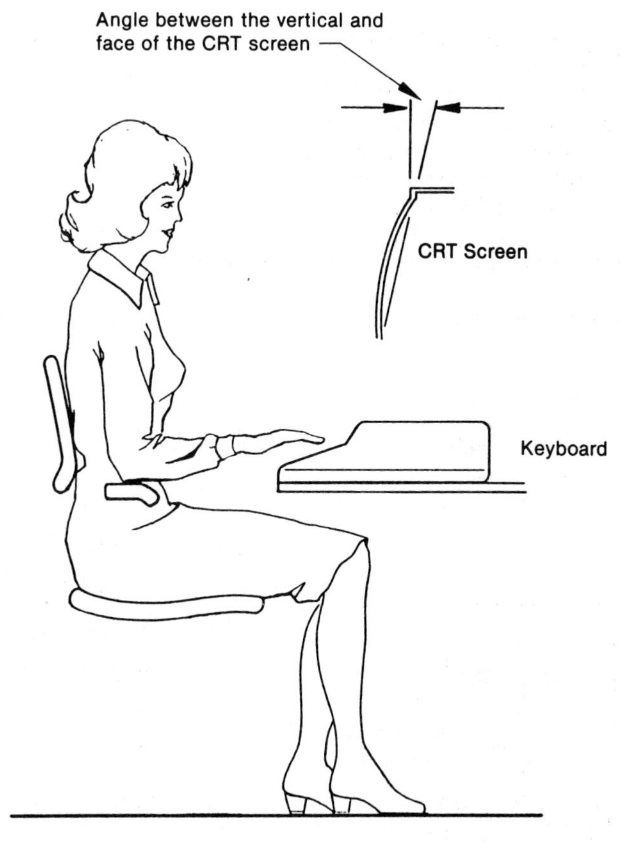

Figure 50. Screen orientation (tilt) for VDU workstation.

Chair

Relevant dimensions for chair design are presented in this section (see Figure 51).

Guidelines	Research Support	Source
1. The chair design should allow the user to maintain the following posture: knees flexed at an angle ≥90° , elbows flexed at an angle ≥90°, and torso at an angle slightly greater than 90° (100° to 155°).	Y	146, 23
2. Seat height should be adjustable from 35 to 55 cm (14 to 22 in.).	Y	150, 146, 42, 119, 71, 130
3. When the chair is provided with a footrest, it should be adjustable from 51 to 76 cm (20 to 30 in.), with the footrest a constant 46 cm (18 in.) below the seat.	Y	150, 42
4. Seat width should be 43 to 51 cm (17 to 20 in.).	Y	150, 42
5. Seat depth should be 38 to 46 cm (15 to 18 in.).	Y	150, 42

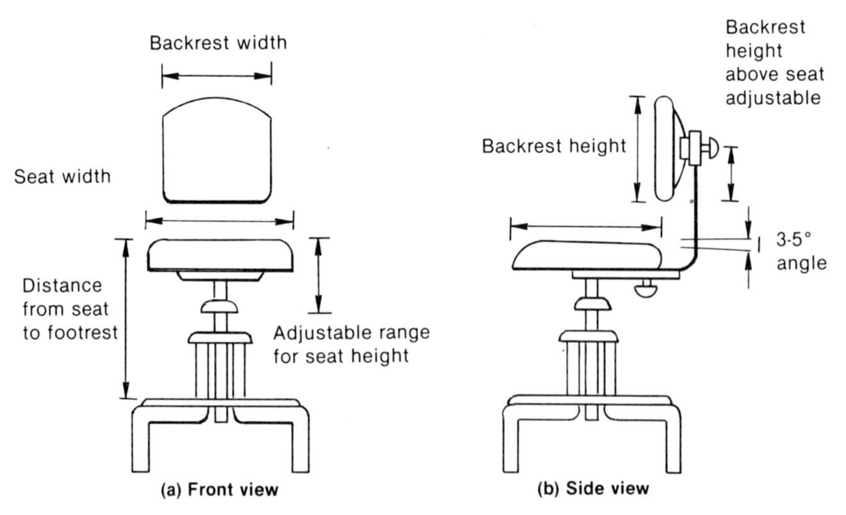

Figure 51. chair dimensions for VDU workstation (adapted from Reference [42]).

6.	Backrest height should be 15 to 23 cm (6 to 9 in.).	Y	42
7.	Backrest width should be 30 to 36 cm (12 to 14 in.).	Y	42
8.	The seat cushion should be at least 1-in. thick.	Y	146
9.	The armrests should be 5 cm (2 in.) wide, 20 cm (8 in.) long, and 19 to 28 cm (7.5 to 11 in.) above the compressed sitting surface. (Swing-away if appropriate.)	Y	146

Comment. In addition to the hard data provided above, consideration should be given to soft issues such as: (a) the chair shape should allow freedom and ability to change posture at frequent intervals, providing appropriate support in these alternate postures; (b) the chair adjustments should be easily manipulated; (c) the seats should be upholstered to reduce sweating; (d) the chair should have casters and provide a swivel feature to allow easy positioning; (e) the composition of the seat surface should not be too hard or too soft; and (f) the chair should be stable on the floor. There is an unconventional approach to seating which suggests an approximate 20° forward seat angle because it imposes much less strain on the neck and back. Only time will tell whether this approach becomes popular.

In recent years, the conventional approach to seating design is being placed under a renewed scrutiny. Guideline 1 dictates the ideal postural considerations that should be followed for achieving optimal seating criteria (e.g., knees flexed at an angle 90 degrees, elbows flexed at an angle of 90 degrees, etc.). Karl Kroemer [70] takes issue with this approach. In an article, published by the Human Factors Society, Kroemer advocates "Free Posture," where the user should not be confined to the rigid requirements of upright 90 degree angles. In the same article, Kroemer states, "I guess there is a revolution going on ...The difference is that you no longer assume that you have to sit upright and have your forearms horizontal. This has brought about a whole new seat design concept. Upright sitting is dead as far as I'm concerned." The authors conclude that, in light of new evidence, more workstations adhere to the free posture philosophy. This approach, that allows users to conform to postures that fit their own individual needs, is of particular significance to operators in a process control environment. In this case comfort, as determined by the individual, is of utmost importance for maintaining a sustained vigilance while monitoring the status of a plant.

Hardcopy Printer

The implementation of a printer in process control often serves to provide a historical record of all parameters monitored by the computer as well as real-time display, whether or not the data are presented on CRT screens. Printers, until recently, have received little attention from the human factors discipline. This is partially due to the overshadowed emphasis on the primary computer interfaces (i.e., VDU and workstation). There is a new awareness to address printer design, since it has been recognized that the printer also comprises a principal interface between the computer and the operator. As a result, the following guidelines should be considered for enhancement of the printer interface.

	Guidelines	Research Support	Source
	Hardware and Physical Characteristics		
1.	Hard-finish, matt paper should be used to avoid smudged copy and glare.	L	150
2.	There should be a positive indication of the remaining supply of recording materials.	L	150
3.	Instructions for reloading paper, ribbon, ink, etc. should appear on an instruction plate attached to the printer.	L	150
4.	Printers should be part of the process computer system and be located in the primary operating area.	L	150
5.	Control room printers should provide the capability to record alarm data, trend data, and plant status data.	L	150

6.	The system should, if possible, be designed to provide hard copy of any page appearing on the CRT at the request of the operator.	L	150
7.	Printer operation should not alter screen content.	L	150
	Operator Information		
8.	A takeup device for printed materials should be provided which requires little or no operator attention and which has a capacity at least equal to the feed supply.	L	150
9.	It should be possible to annotate the print copy while it is still in the machine.	L	150
10.	The operator should always be able to read the most recently printed line.	L	150
11.	Printed material should have an adequate contrast ratio to ensure easy operator reading.	L	150
12.	When the printer is down during reloading, data and information which would normally be printed must not be lost.	L	150
13.	The recorded matter should not be obscured, masked, or otherwise hidden in a manner which prevents direct reading of the material.	L	150
14.	If the copy will be printed remote to the operator, a print confirmation or denial message should be displayed.	L	150
15.	Printed information should be presented in a directly usable form with minimal requirements for decoding, transposing, and interpolating.	L	150
16.	Printer used for recording trend data, computer alarms, and critical status information should have a high-speed printing capability of at least 300 lines a minute to permit printer output to keep up with computer output.	L	150

Comment. Perhaps the most important consideration for a printer is that it should be assessable. Request for hardcopy commands should be simple and user-friendly. In many cases, it might also be advantageous to specify a graphics print capability and flexibility for printing out data on standard 8-1/2 x 11 or legal-size paper.

Health and Safety

The general features of mechanical, electrical, and radiation safety of VDUs are addressed below.

	Guidelines	Research Support	Source
1.	The VDT should be provided with implosion safeguards.[a]	Y	23
2.	No fans, voltage, gears, or belts should be accessible to the user's fingers or body.	Y	18
3.	No parts hotter than 140°F should be accessible to the user.	Y	18
4.	Physical barriers and warnings should be provided for all live parts.	Y	18
5.	Barriers should prevent hardware and small items from falling into areas of high voltage.	Y	18
6.	All exposed corners and sharp edges should be smooth and rounded.	Y	18
7.	Reviews of the literature (on radiation safety) and new surveys to attempt to measure radiation led to the conclusion that there is no radiation hazard for VDT operators and further routine surveys are not necessary.[b]	L	18

[a] Because the VDT operator sits in close proximity to the VDT, it is usual to provide some form of protection for the viewer in the event of CRT breakage. Under normal office circumstances, the risks of such a failure are slight but in some working environments where VDTs have been introduced, e.g.,

factory floor environments, the risks of damage and breakage may be higher [23].

b Although the authors feel the statute to be premature, it has been decided in New York that there is a relationship between working with a VDU and negative health effects in pregnant women. The most recent of these statutes suggests that the pregnant women work no more than 20 hours per week in front of the VDU [75]. Apparently more research is needed before the myriad of radiological health issues associated with VDU usage can be resolved.

Comment. A full treatment of all safety features associated with VDUs is beyond the scope of this document. However, the general guidelines cited above should be considered as the top-level basics when performing a safety review of VDUs.

Environmental Factors

It seems that the most overlooked system during the design and construction of a facility, yet the most complained about once a facility is occupied, is the environmental system (lighting, heating, ventilation, air conditioning, etc.). In the process control industry, these set of systems are often referred to as the HVAC–heating, ventilation, and air conditioning systems. Combining all those factors to produce a functional and comfortable environment requires a large number of considerations. The guidelines presented below are rudimentary to ensure that conditions are acceptable. The blending of these conditions in just the proper ratio will ensure that acceptable conditions can be achieved. Here, as in the anthropometric section, adjustability needs to be stressed. Without the ability to, for example, direct air flow to one's front side rather than back and to adjust illumination levels, a satisfactory situation is unattainable. The guidelines discussed in this section are

- Background noise
- Temperature and humidity
- Lighting
- Ventilation
- Static electricity

Background Noise

Background noise is the maximum ambient noise level (generated by HVAC fans, transformers, lights, etc.) which allows communication with a normal or slightly raised voice level.

Guidelines	Research Support	Source
1. Ambient noise level should be less than 70 dB(A) (less than 65 dB(A) preferred) in routine task areas and less than 55 dB(A) in task areas requiring a high level of concentration.[a]	Y	150, 23, 45, 18
2. Ambient noise should be free from high frequency tones (8000 Hz) and external (or extraneous), high-noise-level equipment.	Y	23, 45

[a] Background noise should not impair verbal communication between any two points in the primary operating area. Verbal communications between these points should be intelligible using normal or slightly raised voice levels [150].

Comment. The prime consideration for these guidelines is allowing for the ability to communicate under the worst conditions which can be encountered over the longest distances to be encountered (alarms actuated but not silenced and two operators at the opposite ends of the control room).

Temperature and Humidity

The ranges of temperature and humidity combinations which conform to the comfort standard recommended by the American Society of Heating, Refrigeration and Air Conditioning Engineers (ASHRAE) [2] for lightly clothed, sedentary individuals in spaces with low air movement (less than 45 fpm) are listed in the following guidelines.

Guidelines	Research Support	Source
1. Ambient temperature should be maintained from 18° to 29.5° K (65° to 85°F), with 21° to 27° K (70° to 80° F) preferred.	Y	150, 146, 23, 18, 2
2. Relative humidity should be from 20 to 60%.[a]	Y	150, 146, 23, 18, 2
3. There should be no more than a 10°F difference between head and floor level.	Y	150

[a] Reference [146] recommends that approximately 45% relative humidity should be provided at 21°C (70°F). This value should decrease with rising temperature but remain above 15%.

Comment. From experience, the equipment (VDU and associated circuitry) drives environmental requirements. As a result, the requirements are typically more restrictive in terms of range and rate of variations; within that range then are the personnel comfort requirements.

Lighting

Sufficient illumination must be delivered to the level of the desk top and/or displays so that displays and printed material can be read accurately. Direct and reflected glare must be minimized.

Guidelines	Research Support	Source
1. Workplace illuminance should be from 92 to 927 lx (adjustable), with a mean of 240 lx during the day shift and a mean of 184 lx during night shift.[a]	Y	150, 146, 23, 45, 18, 130, 72

2.	Emergency lighting level should be from 10 to 50 lx.		46

[a] Only Reference [130] makes a distinction between day and night shift, recommending 105 to 927 lx during day shift and 92 to 332 lx during night shift. References [6] and [141] recommend 200 to 700 lx. References [16], [22], [30], and [40] recommend 300 to 540 lx. All references recommend adjustable illuminance.

Comment. The intent of the guidelines is to provide adequate light for document reading and still allow adequate contrast for easy VDT observation. Minimizing the difference in luminance between surfaces and minimizing glare eases eye fatigue but is difficult to achieve. Coloration of lighting should be noted as well for example, there are documented cases of sodium lighting causing shifts in color spectrum for workers [8]. Many times color-corrected fluorescent lighting can be a workable arrangement.

Ventilation

Ventilation is the quantity, quality, and velocity of air introduced to the conditioned space.

	Guidelines	Research Support	Source
1.	Air should be introduced at a minimum rate of 15 cfm per occupant, approximately 2/3 of which should be outside air that is filtered to remove hazardous or irritating particles.[a]	Y	150, 146, 23, 46
2.	Air velocity should not exceed 45 fpm measured at operator head level and should not produce a noticeable draft.[b]	Y	150, 146, 23, 46

[a] Reference [150] recommends a minimum of 0.85 m^3 (30 ft^3) per minute per occupant.

[b] Reference [23] recommends a maximum of 0.1 m/s (20 fpm). Reference [150] recommends a maximum of 30 m (100 ft) per minute, with less than 20 m (65 ft) per minute preferred.

Comment. With the type and quantity of electronic equipment normally included in a control room, the air quantity requirements will never be determined by personnel needs but rather by equipment heat removal needs. Additionally, an outside air requirement of 0.75 to 2 cfm/ft^2 of floor space is recommended for an office environment (which a control room essentially is) [2]. Additionally, and most important, the direction of air flow should be toward the operators front (most preferable) or the operators side (less preferable), but never toward his back. This is also stressed in Reference [2] but overlooked in the cited sources.

Static Electricity

Static electricity is electricity contained or produced by charged bodies.

	Guidelines	Research Support	Source
1.	Relative humidity should be maintained at 40 ± 10% and should not be allowed to fall below 20%.	Y	46, 2
2.	An earth line should be provided between each VDT and the main system earth connection and carpeting material with a copper wire interweave.	Y	23

Comment. Because of the sensitivity of VDUs to static electricity, both passive (humidity control) and active (grounding pads and conducting systems) static electricity control systems should be provided.

IX

BENEFITS FROM A HUMAN FACTORS ASSESSMENT AND CORRESPONDING METHODS OF EVALUATION

The benefits of a human factors effort have been well described by Meister [89]: "The goal of [Human Factors] is, therefore, to optimize the design of equipment from the standpoint of the equipment user so that his efficiency will be at its greatest. In a nutshell it [Human Factors] attempts to 'tailor' equipment to the capabilities and limitations of the user." A full human factors assessment aids in satisfying the objective of an effective human factors program: (a) improved human performance as shown by increased speed, accuracy, and safety and less energy expenditure and fatigue; (b) less training and reduced training costs; (c) improved use of manpower through minimizing the need for special skills and aptitudes; (d) reduced loss of time and equipment as accidents due to human errors are minimized; and (e) improved comfort and acceptance by the user/operator. Human factors engineering is concerned with improving the productivity of the operator by taking into account human characteristics in designing systems [67]. The payoff is reduced risk and a lower probability of human error. The complexity of modern systems has diminished older, laisez faire attitudes about the human interface. Typical responses such as "the operator is highly adaptable to a poorly designed piece of equipment and has the capability to just muddle through" are rapidly losing ground. Why should millions of dollars be devoted to designing for equipment reliability without considering designing for the reliability of the human who must operate the system?

The purpose of this chapter is to provide a simple resource for evaluating emerging or existing computerized operator support systems against the various guidelines provided in this text. Prior to presenting a table of guidelines and evaluation methods, the reader is invited to review a brief discussion of the various methods available.

There are three general classifications of evaluation techniques. These are checklists, multidimensional rating scales, and performance measurement. Each of these techniques is discussed briefly. More detailed information on these techniques may be found in Kerlinger [78], and limitations of the various assessment methods are well summarized in Guilford [62].

Checklists

Checklists are a primary tool for the evaluation of CRT displays. Checklists have been generated by innumerable institutions and are constantly under revision due to our rapidly increasing knowledge base about CRT displays. A checklist may be used by individuals or administered to a group. If administered to a group, it is subject to all the same constraints placed on any survey regarding (a) attention to sample size and (b) representativeness of the item and population sample. As a whole, the guidelines presented in this text represent the basis for comprehensive checklists. For the reader's convenience, a checklist is available as Appendix A of this document.

Checklists are an excellent means of ensuring compliance with known factors that can affect user performance and are easily implemented through simple observations and linear type measurements. They make up an important component of the overall evaluation of a CRT display. They may be used in combination with a walk-through/talk-through or survey instrument.

Multidimensional Rating Scales (MRS)

Multidimensional rating scales (MRS) are important tools to the evaluator of CRT displays. These scales provide the added component of user acceptance to the overall evaluative process. Although it has been repeatedly shown that user preference does not necessarily correlate with user performance, it is undeniably important that users like the system so that they will use the system. The combination of a checklist followed by the use of a MRS gives added power to the total evaluation of a CRT display. Rating scales may be numeric, graphic, or forced choice. They depend upon human judgment and are popular, perhaps second only to checklists. They are slightly more difficult to construct, tend to present the reviewer with bipolar adjectives against which he or she rates the system under analysis. It should also be noted that rating scales can be subject to bias from halo effects, logical errors, and extreme response tendencies [62]. For the most part, it is best when applying an MRS to adapt an existing rating instrument whose reliability and validity have been established.

Performance Measurement

After a system has been developed, a critical component of the overall evaluation process is the performance evaluation. This step is the actual testing of users with the system to determine if the users can, in fact, perform as desired. It is a technique that can also be used to determine which of several alternate designs might be implemented. This performance testing is the final proof of the display after all the guidelines have been applied. In general, practitioners will wish to apply a quasi-experimental design in which users or potential users are presented with different scenarios that exercise the capability of the system. This design can extend from starting up the system to reviewing operator response when system power is lost.

The degree to which a simulation may be used prior to casting the final design in concrete is a matter of engineering judgment. If a full-scope simulation can be used to fast prototype an initial design and allow for quick change modifications, then there are cost savings to be gained in terms of both reduced errors and enhanced operator performance.

With the application of all these methods, one will have a complete evaluation of a CRT workstation. These methods in combination will account for the greatest amount of the potential variability of performance on the part of the user. The confidence in the performance of the system will be reduced if all methods are not applied in the evaluative process.

Table 14 was constructed to parallel the guidelines given in this text. It is organized similarly for ease of use. The types of measurement should be self-explanatory.

TABLE 14. METHODS FOR EVALUTION AND ASSESSMENT OF UCI GUIDELINES

	Visual Observation	Operate	User Interiew	Direct Measurement	Manufacturer Specification	Task Analysis	Auditory Observation
Hardware Aspects							
Flicker	X						
Contrast ratio				X			
Display luminance				X			
Phosphor	X						
Glare	X						
Screen resolution	X						
Screen Structures and Content							
Cursor	X	X					
Text	X		X				
Labels	X						
Messages	X		X				
Abbreviations	X		X				
Error statements	X		X				
Nontextual messages	X						
Data display	X						
Data entry	X						
Instructions	X		X				
Characteristics of Alphanumeric Characters							
Font or style	X						
Character size and proportion	X			X Ruler			
Character case	X						

TABLE 14.(*continued*)

	Visual Observation	Operate	User Interiew	Direct Measurement	Manufacturer Specification	Task Analysis	Auditory Observation
Screen Organization and Layout							
Screen size	X						
Grouping			X			X	
Display Density	X			X			
Display Partitioning/windows	X						
Frame specifications	X						
Interframe considerations-paging and scrolling	X						
Interframe considerations-windowing				X			
Visual Coding Dimensions							
Color	X		X	X verify color			
Geometric shape	X		X	X			
Pictorial	X		X				
Magnitude	X		X	X			
Enhancement Coding Dimensions							
Brightness	X		X				
Blink	X		X	X watch			
Image reversal	X		X				
Auditory			X	X			X
Voice			X	X			X
Audio–visual	X		X	X			X

TABLE 14.(*continued*)

	Visual Observation	Operate	User Interiew	Direct Measurement	Manufacturer Specification	Task Analysis	Auditory Observation
Other techniques	X		X				
Dynamic Display							
Display motion	X		X				
Digital counters	X			X watch			
Information Formats							
Analog	X			X			
Digital	X			X			
Binary indicator	X						
Bar/column charts	X						
Band charts	X		X				
Linear profile	X		X				
Circular profile	X		X				
Single value line chart	X		X				
Trend plot	X		X				
Mimic display	X		X				
Keyboard Layout							
Keystroke feedback		X	X		X		
Key actuation force		X	X	X	X		
Key rollover		X	X	X	X		
Key travel (displacement)		X		X	X		
Key color/labeling	X						
Key dimension/spacing	X			X			
Keyboard slope				X			

TABLE 14.(*continued*)

	Visual Observation	Operate	User Interiew	Direct Measurement	Manufacturer Specification	Task Analysis	Auditory Observation
Keyboard thickness				X			
Special function keys	X		X				
Soft programmable keys	X	X	X				
Numeric keypad	X						
Alternate Input Devices							
Light pens	X	X	X	X			
Joysticks	X	X	X	X			
Tracker ball	X	X	X	X			
Grid-and-stylus-devices	X	X	X	X			
Free–moving X–Y controller (mouse)	X	X	X	X			
Automatic speech recongnition (voice input device)	X	X	X	X			X
Touch screen input	X	X	X	X			X
User Dialogue							
Question and answer	X	X					
Form filling	X	X					
Menu design	X	X	X				
Command language	X						
Query language	X						
Natural language	X						
Expert Systems	X	X	X			X	

TABLE 14.(*continued*)

	Visual Observation	Operate	User Interiew	Direct Measurement	Manufacturer Specification	Task Analysis	Auditory Observation
System Feedback							
Display update rate			X		X		
Response time	X	X	X	X	X		
System status indication	X						
Routine status information	X						
Performance/Job aids	X		X			X	
Software Security							
Data Protection/Data Security		X	X	X			
Anthropometrics							
Keyboard base				X			
Working level				X			
Keyboard home row				X			
Screen				X			
Viewing distance				X			
Footrest				X			
Reach envelope				X			
Position and movement of the head				X			
Leg, knee, and foot room				X			
Screen orientation				X			
Chair				X			
Hardcopy printer	X						
Health and safety	X						

TABLE 14.(*continued*)

	Visual Observation	Operate	User Interiew	Direct Measurement	Manufacturer Specification	Task Analysis	Auditory Observation
Environmental Factors							
Background noise		X		X			X
Temperature and humidity				X			
Lighting				X			
Ventilation	X						
	Olfactory						
Static electricity	X			x			

X

CONCLUSIONS AND SUMMARY

From the perspective of the plant operator, the rigorous application of systems engineering principles, design guidelines, and assessment/evaluation techniques can ultimately ensure the implementation of a user interface that will be both acceptable and effective.

The guidelines that are presented review many lessons learned and research findings that, in the authors' experiences, have made a design successful. These lessons learned are summarized in the following sections. It is hoped that the system designer will be assisted in identifying the myriad of pitfalls that beset every large-scale design effort. This chapter focuses on both lessons from the past, and for future interface design. We will briefly examine where human factors engineering in process control is going, what has been done, and what still remains to be done.

Lessons and Observations From the Past

Guidelines are Not a Substitute for Testing and Evaluation.

The interactive complexities of human factors engineering guidelines are not well understood. Although the judicious use of published human factors engineering guidelines is a necessary requirement in the design of user-computer interfaces, they are not a substitute for testing the usability of a system.

Use a Variety of Techniques for Testing System Usability.

Without overburdening the user, employ a variety of methods in a test plan. Do not overrely on any single method. The implementation of more than one method is encouraged; however, the test analyst should be aware of the advantages and disadvantages of each method to be applied. For example, the use of three similar methods, such as rating scale, checklist, and multiple choice paper/pencil instruments, is not as effective as a plan that includes rating scale, observation, and structured interview.

Techniques for Constructing Paper and Pencil Usability Evaluation Instruments

(a) Design the response items of a questionnaire in a style that makes them readily understood by the user, e.g., avoid jargon; (b) write simply and clearly; (c) ask discrcet questions; and (d) keep the number of response options to a minimum.

Verify that the response item can be answered within the total context of the software program (e.g., if the use of color coding, screen clutter, and highlighting techniques are significantly different across screens, then either delete the response item or make a specific response item for each screen).

Avoid the use of generic questionnaires if time and resources are available to develop an original one. Generic questionnaires are not sensitive toward finding specific issues and problems that may exist in a particular application. No off-the-shelf questionnaire is an acceptable substitute for constructing an original questionnaire that can effectively be used to address the specific issues of a software interface and match the vocabulary and experience of the raters.

When selecting methods for usability testing, don't underestimate the usefulness of simple observation and structured interview techniques.

Defining the Scope of Software Usability Evaluation

A comprehensive test plan should not be restricted to a single subset of applications or operating modes. Select scenarios, e.g., examples, that test the full extent of the software program. If, for example, the user interface is to be used in the execution of emergency operating procedures during off-normal events, then the test scenarios for simulating this situation should be included in the test plan.

User Guidance

Don't automatically assume that on-line tutorials can teach. Untold resources have been spent on the assumption that embedded training modules can solve all the problems of operator training on a new system.

Remember on-line HELP is only as effective as the quality of the original source material used. A HELP system that furnishes rapid user access to the information in a creative format is worthless if the original material used in the development of the HELP screens is poor.

Know the End User

It is not enough to simply design for the user. More importantly one should design with the user. Users of a system should be active participants as subject matter experts (SMEs) throughout all phases of system development. This approach has the dual purpose of (a) gaining insight to the operations of a system from a hands-on perspective, thus increasing the likelihood of a better design for the user in mind, and (b) enabling the user to buy off on the system after it has been implemented and turned over to them. In knowing the end user, it is also important to differentiate between a true SME and the user's representative [118]. Be sure the SMEs and test participants for whom the system is being developed are selected from the user population, not the user's manager.

Know the System

The system designer who is responsible for the user interface must know what the system will do, how it will be used, and who will operate it. The hope of achieving a usable interface is shattered when systems are designed and procured by engineers with little or no background in plant operations. The system designer cannot be an expert in every phase of plant operation, but he/she can learn the basics by performing system walkdowns of the hardware and asking the operators questions about their job.

Don't Confine the System Design to Hardware/Software

The completed system must be integrated into the overall operational setting. At the early stages of system development, the design team must take into consideration the peripheral issues that indirectly impact the performance of the system. For example, how will the new system affect (a) training development (what type and degree of training support will be required?); (b) procedures/technical specifications (to what extent should the operating procedures be revised?); (c) crew staffing (what should the delineation of operator responsibilities be?); (d) workload (will the complexity of the job warrant additional personnel to be assigned on the shift?); (e) administrative/management controls (how will the system affect the responsibilities and duties between organizations?); and (f) control room environment (will the ambient lighting and background noise levels adversely affect operator performance?).

Don't Become Overly Enamored with New Technologies

When selecting a technology for the system, it is important to observe the watchword to keep it simple. There have been numerous cases in the consumer market when users were found to be completely frustrated and confused by products that employed new and innovative technologies [109]. One does not have to look past their own automobile to see where many of these new technologies failed to keep their original claims [21]. It is questionable, for example, if instrumentation, power spectrums for torque, and RPM curves have added anything to the ordinary driver's ability to navigate down the highway.

Learn When to Freeze a Design and Live with it

When designing a system, it is very easy to lose sight of the constraints of schedule and budget available for completing a project. A design team seldom has the leisure to develop the ideal system. At best, one strives to optimize design within the available resources. All too often a system fails because new requirements are added throughout all the phases of system development. In an attempt to be all things to all people, the system becomes nothing to anyone. After a certain point in the design process, the fine tuning of the system must stop and implementation has to begin.

Remember to Document All Human Factors Recommendations

If a human factors recommendation is worth incorporating into the final design, it's worth writing down. Don't automatically assume that human factors design criteria will be included in the design from a sideline conversation or informal discussion with a team of designers; in most cases, they won't!

Lessons and Observations for the Future Interface design

The State of Human Factors Engineering Guidelines

In an era of rapidly changing technology, there is still a pressing need for more guidelines for user-computer interaction. Although many fundamentals for the hardware side of user-computer interaction are well defined, useful guidelines for software are extremely limited. To date, guidelines for design and development of expert systems, graphical interfaces, and voice communication are almost nonexistent. As technology continues to move forward, those human factors guidelines that do exist become dated and, as a result, less useful to the designer.

Also at issue is a need for rigorous validation studies for many of the guidelines that have already been published. For example, many of the often cited guidelines for CRT displays are still grounded in hardcopy media and expert judgment. In the process control rooms, it is not unusual to see the operator presented with the status of well over a thousand signals. Many hard questions remain in regard to just how this information should be packaged, interpreted, and presented in a format for decision-making.

Even guidelines that are valid and current may not be applicable in all cases to all systems. For example, guidelines for screen structures, such as icons and statistical charts, may be perfectly acceptable for high-resolution color graphic monitors but how do the same guidelines apply to low-resolution displays? From another perspective, are the same guidelines suitable for high-resolution CRT colorgraphics acceptable for electroluminescent displays?

In summary, more work is needed to (a) validate many of the guidelines that do exist and (b) perform research in selected areas for the purpose of generating guidelines that are not yet available.

The State of Human Factors Testing and Evaluation

As pointed out in the previous section, guidelines, though useful, are highly dependent upon each other and their interactive complexities are not well understood. This characteristic of guidelines makes it difficult to conclude that a product has all the attributes of an acceptable interface simply because it satisfies a set of general human factors criteria. In short, without test and evaluation, there is no easy alternative for ensuring that the system meets the attributes of a satisfactory user interface.

Nonetheless, after the initial decision has been made to include a test and evaluation of the user interface, then a valid methodology should be defined. To date, the state of the art for usability testing has not progressed to the point where one can categorically state that "Method X" is the best overall. Even if high-fidelity studies are conducted of human performance using simulators, the dependent measures are often difficult to clarify in a process control setting. The classical measures of errors and time may be useful for assessing performance of word-processing packages, but to what extent do these factors effect the mitigation of a plant safety event? With evolutions taking upwards of six hours, how many of those entry and reading errors would be fully recovered by another operator when a second recheck of the system's status is made?

In a study by Blackman and Gilmore [14], it was noted that the percent of time that plant parameters (e.g., temperature, pressure, level) were operating outside of set point ranges was an inadequate measure of performance. In this study, it turned out that during emergency operations the operators were not concerned with keeping specific parameters within acceptable limits. Instead, they were more concerned with the bigger picture of bringing the plant to a safe shutdown. Measurement problems have been identified with the use of assessing performance based on modeling the decision-making process.

Blackman and Gilmore also investigated the effectiveness of safety parameter display systems (SPDSs) using operator action event trees (OAETs) as methods of scoring operator paths in transient correction. A typical OAET is illustrated in Figure 52. The findings revealed that the accuracy of the solution set was highly dependent upon the operators' own style and philosophy for handling a transient condition. Although the operators' actions are backed up by an array of technical specifications and procedures, their strategies for handling an emergency are still somewhat variable between individuals, as well as amongst crews. If Operator "A" brings the plant to a stable condition without deviating from the operator action event tree,

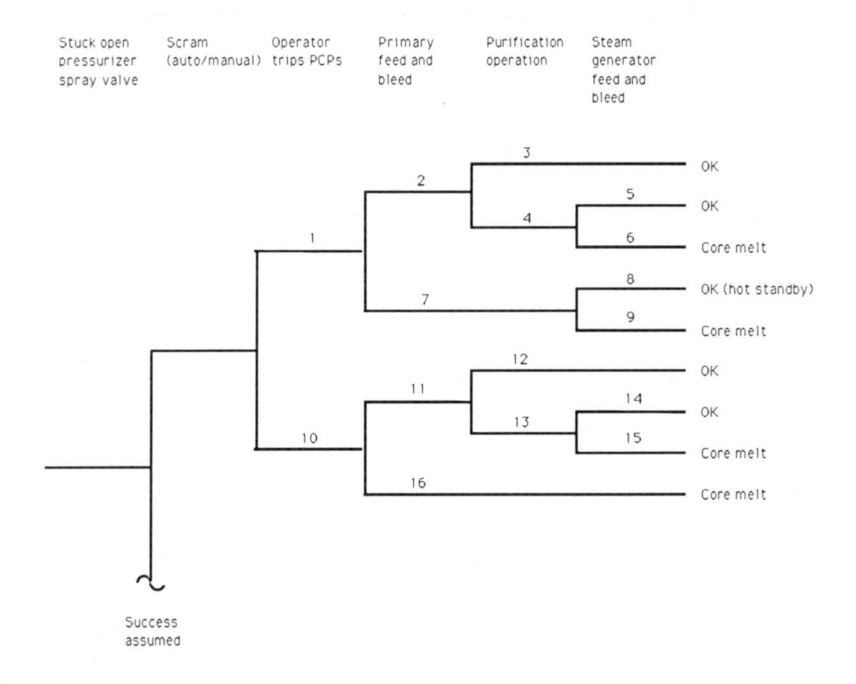

Figure 52. Example of an operator action Event Tree (OAET)

his/her score from the experiment would have been very high. However, if Operator "B" adapts a novel strategy for mitigating the transient that is in violation of the action event tree, his/her score would have been somewhat lower than the score of Operator "A." If both operators saved the plant (the end goal), is one strategy better than another? The problem of finding a suitable method is also complicated by the lack of acceptable operational definitions for terminology such as information quality, screen clutter, and user-friendly dialogue. For example, what are the characteristics of a user-friendly system? How do I test for user-friendliness?

A great deal of work still remains to be done before standardized methods for performing test and evaluations of operator interfaces in process control settings can be established.

In summary, the field of user-computer interaction for process control is still in its infancy. It is the mission of all specialists, from computer science to cognitive psychology, to take an active role in developing guidelines, tools, and methods that will ensure that the systems of the future can truly meet the designers challenge of being truly user-friendly.

FIGURES

TABLES

REFERENCES

1. "American Standard for Human Factors Engineering of Visual Display Terminal Workstations" The Human Factors Society Bulletin (ANSI/HFS 100-1988), Santa Monica, California, February 4, 1988.
2. American Society of Heating, Refrigeration and Air Conditioning Engineers (ASHRAE), Handbook of Fundamentals, New York, 1972.
3. D. G. Alden, Human Factors Principles for Keyboard Design and Operation—A Summary Review, Honeywell Systems and Research Division, Minneapolis, Minnesota, March 26, 1970.
4. C. A. Baker and W. F. Grether, Visual Presentation of Information, WADCTR54-160, Wright Air Development Center, WPAFB, Dayton, Ohio, 1954.
5. R. W. Bailey, Human Performance Engineering: A Guide for System Designers, Englewood Cliffs, New Jersey: Prentice Hall, Inc., 1982.
6. W. W. Banks and M. P. Boone, Nuclear Control Room Annunciators: Problems and Recommendations, NUREG/CR-2147, U.S. NRC, September 1981.
7. W. W. Banks, D. I. Gertman, and R. J. Petersen, Human Engineering Design Consideration for Cathode Ray Tube-Generated Displays, NUREG/CR-2496, U.S. NRC, April 1982.
8. W. W. Banks, D. I. Gertman, and R. L. Sprague, "Shedding Light on Graphic Artists: A Note on High-Pressure Sodium Lighting," Human Factors Society Bulletin, 26 (5), May, 1983.
9. W. W. Banks et al., Human Engineering Design Considerations for Cathode Ray Tube-Generated Displays, 2, NUREG/CR-3003, U.S. Nuclear Regulatory Commission, July 1983.
10. J. E. Barmack and H. W. Sinaiko, Human Factors Problems in Computer Generated Graphic Displays, Report S-234, Institute for Defense Analyses, 1966.

11. C. Berger, "I, Stroke-Width, Form and Horizontal Spacing of Numerals as Determinants of the Threshold of Recognition," Journal of Applied Psychology, 28, 1944, pp. 208-231.

12. C. Berger, "II, Stroke-Width, Form and Horizontal Spacing of Numerals as Determinants of the Threshold of Recognition," Journal of Applied Psychology, 28, 1944, pp. 336-346.

13. H. M. Bowen et al., Optimal Symbols for Radar Displays, Technical Report 2682 (00) (AD 227 014), Office of Naval Research, Washington, D.C., September 1959.

14. H. S. Blackman and W. E. Gilmore, Interactive Simulator Evaluation for CRT Generated Displays, EGG-2308, NUREG/CR-3767, U.S. NRC, December 1984.

15. H. S. Blackman and W. E. Gilmore, "The Analysis of Visual Coding Variables on CRT Generated Displays," Proceedings of the Society for Information Display (SID), 29 (2), Playa Del Rey, California, 1988.

16. B. S. Blanchard and W. J. Fabrycky, Systems Engineering and Analysis, Englewood Cliffs, New Jersey: Prentice-Hall, Inc., 1981.

17. C. R. Brown and D. L. Schaum, "User-Adjusted VDU Parameters," E. Grandjean and E. Vigliani (eds.), Ergonomic Aspects of Visual Display Terminals, Proceedings of the Internation Workshop, Milan, March 1980, London: Taylor & Francis Ltd., 1983.

18. D. M. Brown et al., Human Factors Engineering Criteria for Information Processing Systems, Sunnyvale, California: Lockheed Missiles and Space Co., Inc., June 1981.

19. B. G. Buchanan, et al., "Constructing an Expert System," Building Expert Systems, I, F. Hayes-Roth, D. A. Waterman, and D. B. Lenat (eds.), London: Addison-Wesley Publishing Company, Inc., 1983.

20. A. Burg, "Visual Acuity as Measured by Static and Dynamic Tests: A Comparative Evaluation," Journal of Applied Psychology, 50, 1966, pp. 460-466.

21. “Smart Cars” Business Week, pp. 68-74, June 13, 1988.

22. A. Cakir et al., Anpassung von Bildschirmarbeitsplatzen an die Physische und Funktionsweise des Menschen, April 1978.

23. A. Cakir, D. J. Hart, and T. F. M. Stewart, Visual Display Terminals, Chichester: John Wiley & Sons, 1980.

24. J. R. Carbonell, J. I. Elkind, and R. S. Nickerson, "On the Psychological Importance of Time in a Time-Sharing System," Human Factors, 10, 1968, pp. 135-142.

25. J. M. Carroll and M. B. Rosson, "Usability Specification as a Tool in Iterative Development," Rex H. Hartson (ed.) Advances in Human Computer Interaction, 1, Norwood, New Jersey: Ablex Publishing Corp., 1985.

26. R. C. Casperson, "The Visual Discrimination of Geometric Forms," Journal of Experimental Psychology, 40, 1950, pp. 668-681.

27. C. Chapanis and R. Halsey, "Absolute Judgments of Spectrum Colors," Journal of Psychology, 42, 1956, p. 99.

28. R. E. Christ, "Review and Analysis of Color Coding Research for Visual Displays," Human Factors, 17, 1975, pp. 542-570.

29. D. S. Ciccone, M. G. Samet, and J. B. Channon, A Framework for the Development of Improved Tactical Symbology, Technical Report 403, U.S. Army Research Institute for the Behavioral and Social Sciences, August 1971.

30. J. Cohen and A. J. Dinnerstein, Flash Rate as a Visual Coding Dimension for Information, WADCTR57-64, Wright Air Development Center, WPAFB, Ohio, 1958.

31. C. Conover, The Amount of Information in the Absolute Judgment of Munsell Hues, WADC-TN58-262, Wright Air Development Center, WPAFB, Dayton, Ohio, 1959.

32. E. R. F. W. Crossman, "Automation and Skill," The Human Operator in Process Control, E. Edwards and F. P. Lees (eds.), London: Taylor and Francis Ltd. 1974.

33. R. Dallimonti, "Challenge for the `80's: Making Man-Machine Interfaces More Effective," Control Engineering, 29 (2), January 1982, pp. 26-30.

34. M. M. Danchak, "The Content of Process Control Alarm Displays," Paper presented at the Instrument Society of America (ISA), International Conference and Exhibit, Houston, Texas, October 20-23, 1980.

35. M. M. Danchak, Techniques for Displaying Multivariate Data on Cathode Ray Tubes with Applications to Nuclear Process Control, NUREG/CR-1994, U.S. NRC, April 1981.

36. M. M. Danchak, "Effective CRT Display Creation for Power Plant Applications," Proceedings of the Nineteenth International ISA Power Instrumentation Symposium, San Fransico, California, May 9-12, 1976.

37. M. M. Danchak, Alarms Within Advanced Display Systems: Alternative and Performance Measures, NUREG/CR-2776, U.S. NRC, September 1982.

38. C. J. Davis, Radar Symbology Studies Leading to Standardization: II, Discrimination in Mixed Displays, Technical Memorandum 5-69 (AD688 125), Human Engineering Laboratories, Aberdeen Research and Development Center, Aberdeen Proving Grounds, Maryland, March 1969.

39. A. B. Dickinson, "Remember Operator Needs When Selecting CRT Displays," Instruments and Control Systems, 52 (4), pp. 37-41, March 1979.

40. H. Dreyfus, Symbol Sourcebook, New York: McGraw-Hill, 1972.

41. DIN Standard 66 234, Characteristic Values for the Adaptation of Work Stations With Fluorescent Screens to Humans Deutsches Institut fur Normung. (Draft).

42. Eastman Kodak Co., Ergonomic Design for People at Work, 1, Belmont, California: Lifetime Learning Publications, 1983.

43. E. Edwards and F. P. Lees, The Human Operator in Process Control, London: Taylor & Francis Ltd., 1974.

44. D. R. Eike and C. C. Baker, "Status of Update of DOD-HDBK-761 Human Engineering Guidelines for Management Information Systems (HEGMIS)," paper presented at DOD-HFE-TAG, UCI Working Group Meeting, Vail, Colorado, July 21, 1987.

45. Electric Power Research Institute (EPRI), Computer-Generated Display System Guidelines, 1 and 2, EPRI NP-3701, Interim Report, Palo Alto, California, September 1984.

46. Electric Power Research Institute, Human Factors Guide for Nuclear Power Plant Control Room Development, EPRI NP-3659, Final Report, Palo Alto, California, August 1984.

47. S. E. Engel and R. E. Granda, Guidelines for Man/Display Interfaces, Technical Report TR 00.2720, IBM, Poughkeepsie, New_York, December 1975.

48. W. K. English, D. C. Engelbart, and M. L. Berman, "Display Selection Techniques for Text Manipulation," IEEE Transactions on Human Factors in Electronics, HFE-8, 1967, pp. 5-15.

49. T. D. Faulkner and R. A. Day, "The Maximum Functional Reach for the Female Operator," AIIE Transactions, 2, 1970, pp. 126-131.

50. J. D. Foley and V. L. Wallace, "The Art of Natural Graphic Man-Machine Conversation," Proceedings of the IEEE, 62 (4), 1974, pp. 462-471.

51. J. D. Foley, V. L. Wallace, and P. Chan, The Human Factors of Graphics Interaction: Tasks and Techniques, Report No. GWU-IIST-81-3, George Washington University, Washington, D.C., January 1981.

52. W. O. Galitz, "Debut II—The CNA Data Entry Utility," Proceedings of the 1979 Human Factors Society Meeting, Boston, Massachusetts, October 29-November 1, 1979, pp. 50-54.

53. W. O. Galitz, Handbook of Screen Format Design, Wellesley, Massachusetts: Q.E.D. Information Science, 1981.

54. W. O. Galitz, Human Factors in Office Automation, Life Office Management Association, Inc., Atlanta, Georgia, 1980.

55. D. I. Gertman, L. R. Klinestiver, and J. T. McConville, Human Engineering Analysis Assessment of Disproportionate Anthropometrics and Predictions for the Naval Aviator Population, unpublished Technical Report (available from David I. Gertman, INEL/EG&G Idaho, Inc., Idaho Falls, Idaho), U.S. Naval Air Systems Command, October 1987.

56. L. P. Goodstein, "Functional Alarming and Information Retrieval," Riso-M 2511, Riso_National Laboratory, Halden, Norway, August 1985.

57. N. C. Goodwin, "Cursor Positioning on an Electronic Display Using Lightpen, Lightgun, or Keyboard for Three Basic Tasks," Human Factors, 17, 1975, pp. 289-295.

58. E. L. Gorrel, A Human Engineering Specification for Legibility of Alphanumeric Symbology on Video Monitor Displays, Technical Report No. 78746, Defense and Civil Institute of Environmental Medicine (DCIEM), December 1978.

59. Graphical Symbols for Process Flow Diagrams in the Petroleum and Chemical Industries, ANSI Y32.11-1961, American Society of Mechanical Engineers (ASME), New York, 1961 (Reapproved in 1985).

60. E. Grandjean and E. Vigliani (eds.), "Ergonomic Aspects of Visual Display Terminals," Proceedings of the International Workshop, Milan, March 1980, London: Taylor & Francis Ltd., 1983.

61. J. P. Gould et al., "Why is Reading Slower From CRT Displays Than From Paper," Proceedings of the Human Factors Society—30th Annual Meeting, Dayton, Ohio, September 29-October 3, 1986.

62. J. P. Guilford, Psychometric Methods, Second Edition, New York: McGraw-Hill Book Co., 1954.

63. P. Harmon and D. King, Expert Systems, Artificial Intelligence in Business, New York: John Wiley & Sons, Inc., 1985.

64. P. W. Hemingway, A. L. Kubala, and G. D. Chastain, Study of Symbology for Automated Graphic Displays, ARI Technical Report TR-79 A18, U.S. Army Research Institute for the Behavioral and Social Sciences, Alexandria, Virginia, May 1979.

65. R. J. Hornick, "Dreams—Design and Destiny", Presented at the Human Factors Society's Thirtieth Annual Meeting, Dayton, Ohio, September 29-October 3, 1986.

66. W. C. Howell and C. L. Kraft, Size, Blur, and Contrast as Variables Affecting the Legibility of Alphanumeric Symbols on Radar-Type

Displays, Wright-Patterson Air Development Center Technical Report, September 1959, pp. 59-536.

67. R. D. Huchingson, New Horizons for Human Factors in Design, New York: McGraw-Hill Book Co., 1981.

68. R. D. Huchingson, L. J. Lampen, and R. J. Koppa, "Keyboard Design and Operation: A Literature Review" Report No. RF7020-1F, Texas Transportation Institute, College Station, Texas.

69. Human Factors for Designers of Naval Equipment, 2, British Royal Navy (BRN) Chapters VI and VII, London, England: HMRN, 1971.

70. "Ergonomics Today: Interviews on the Field of Ergonomics," Human Factors Society Bulletin, 31 (9), September 1988.

71. W. Hunting, T. H. Laubli, and E. Grandjean, "Constrained Postures of VDU Operators," E. Grandjean and E. Vigliani (eds.), "Ergonomic Aspects of Visual Display Terminals," Proceedings of the International Workshop, Milan, March 1980, London: Taylor & Francis Ltd., 1983.

72. J. E. Kaufman and J. F. Christensen (eds.), IES Lighting Handbook, Fifth Edition, New York: Illuminating Engineering Society, 1972.

73. K. H. E. Kroemer, "Seating in Plant and Office," American Industrial Hygiene Association Journal, 32 (10) 1971, pp. 633-652.

74. IBM Human Factors Center, Human Factors of Workstations With Visual Displays, Third Edition, G320-6102-2, San Jose, California: International Business Machines Corp., January 1984.

75. The Institute of Electrical and Electronics Engineers, Inc., Controversial Standards Legislated for VDTs, 12 (8), August 1988.

76. Instrument Society of America (ISA), Graphic Symbols for Process Displays, ANSI/ISA-S5.5-1985, approved February 3, 1986.

77. G. W. Irving et al., ODA Pilot Study II: Selection of an Interactive Graphics Control Device for Continuous Subjective Functions

Applications (Report No. 215-2), Integrated Sciences Corp., Santa Monica, California, April 1976.

78. F. N. Kerlinger, Foundations of Behavioral Research, Second Edition, New York: Holt, Rinehart, and Winston, Inc., 1973.

79. L. E. King and G. Tierney, Legibility: Symbols Versus Word Highway Signs, Highway Research Information Service No. 54-225961, West Virginia University, Morgantown, West Virginia, 1970.

80. J. W. Kling and L. A. Riggs (eds.), Woodworth and Schlosberg's Experimental Psychology, Third Edition, Volume 1: Sensation and Perception, New York: Holt, Rinehart, and Winston, Inc., 1972.

81. B. G. Knapp, F. L. Moses, and L. H. Gellman, Information Highlighting on Complex Displays, Directions in Human/Computer Interaction, A. Badre and B. Shneidnerman (eds.) Norwood, New Jersey: Ablex Publishing Corp., 1982, pp. 195-215.

82. C. G. Koch and T. R. Edman, "Evaluating Options for CRT Display of Process Trend Data," Proceedings Human Factors Society 26th Annual Meeting, Seattle, Washington, 1982, pp. 49-53.

83. P. A. Kolers, R. L. Duchnicky, and D. C. Ferguson, "Eye Measurement of Readability of CRT Displays," Human Factors, 23, 5, pp. 517-527.

84. C. Lu, "Computer Pointing Devices: Living With Mice," High Technology, January 1984, pp. 61-67.

85. J. Mahaffey, private communication, Georgia Institute of Technology, February 23, 1983, Human Engineering Design Considerations for Cathode Ray Tube-Generated Displays, II, NUREG/CR-3003, U.S. Nuclear Regulatory Commission, Washington, D.C..

86. W. C. Mair, R. Wood, and W. Davis, Computer Control and Audit, Fourth printing, Altamonte Springs, Florida: Institute of International Auditors, 1982.

87. J. Martin, Design of Man-Computer Dialogues, Englewood Cliffs, New Jersey: Prentice-Hall, 1973.

88. C. M. Mavrides, "Codability of Polygon Patterns," Perceptual and Motor Skills, 37, 1973, pp. 343-347.

89. D. Meister, Human Factors: Theory and Practice, New York: Wiley-Interscience, a Division of John Wiley & Sons, Inc., 1971.

90. E. J. McCormick, Human Factors in Engineering and Design, Fourth Edition, New York: McGraw-Hill Book Co., 1976.

91. E. J. McCormick and M. S. Sanders, Human Factors in Engineering and Design, Second Edition, New York: McGraw-Hill Book Co., 1982.

92. Microsoft Corporation, Microsoft Windows, User's Guide, Version 1.03, Redmond, Washington, 1987.

93. V. Mitchell-Bishop, "Human Operator Performance in Computer-Based Message Switching Systems: A Case Study," Proceedings of the 1979 Human Factors Society Meeting, Boston, Massachusetts, October 29-November 1, 1979, pp. 45-49.

94. R. B. Miller, "Response Time in Man-Computer Conversational Transactions," AFIPS Conference Proceedings (Fall Joint Computer Conference, 1968), Washington, D.C.: Thompson Book Company, 1968.

95. J. W. Miller and E. Ludvigh, "The Effect of Relative Motion on Visual Acuity," Survey Ophthalmology, 7, 1962, pp. 83-116.

96. W. Miller and T. W. Suther, III, "Display Anthropometrics: Preferred Height and Angle Settings of CRT and Keyboard," Human Factors, 25 (4), 1983, pp. 401-408.

97. R. Modley, "World Language Without Words," Communication, Autumn, 1974.

98. P. A. Moore, R. R. Pierce, and D. Shamonsky, How to Design an Effective Graphics Presentation, 17, Harvard Graduate School of Design, Cambridge, Massachusetts, 1981.

99. F. A. Muckler (ed.), Human Factors Review: 1984, The Human Factors Society, Inc., Santa Monica, California, 1984.

100. A. H. Munsell, Book of Color, Baltimore: Munsell Color Book Co., 1942.

101. B. B. Murdock, "The Serial Effect of Free Recall," Journal of Experimental Psychology, 64, 1962, pp. 482-488.

102. W. E. Murray et al., "A Radiation and Industrial Hygiene Survey of Video Display Terminal Operations," Human Factors, 23 (4), 1981, pp. 413-420.

103. P. Muter et al., "Extended Reading of Continuous Text on Television Screens," Human Factors, 24, (5), October 1982, pp. 501-508.

104. W. R. Nelson, "Expert Systems for Diagnosis and Treatment of Nuclear Reactor Accidents," paper presented at the Workshop on Artificial Intelligence Application to Nuclear Power Technology, Electric Power Research Institute, May 8-9, 1984, Palo Alto, California.

105. W. R. Nelson and H. S. Blackman, Response Tree Evaluation: Experimental Assessment of an Expert System for Nuclear Reactor Operators, NUREG/CR-4272, U.S. NRC, September 1985.

106. W. M. Newman and R. F. Sproull, Principles of Interactive Computer Graphics, New York: McGraw-Hill, 1979.

107. H. H. Ng and S. J. Puchkoff, "Touchsensitive Screens Ensure a User Friendly Interface," Computer Design, The Magazine of Computer Based Systems, August 1981.

108. (NISO), National Information Standards Organization, Proposed American National Standard for Information Sciences—Common Command Language for Online Interactive Information Retrieval (Z39.58-198x), circulated for comment July 1, 1987 to September 30, 1987.

109. D. A. Norman, The Psychology of Everyday Things, New York: Basic Books, Inc., 1988.

110. Nuclear Software Services, Inc., A Primer on Colorgraphic Display Systems for Nuclear Power Plants, NSAC-45, Nuclear Safety Analysis Center, Operated by Electric Power Research Institute, Palo Alto, California, April 1982.

111. J. A. Odom, Applying Manual Controls and Displays: A Practical Guide to Panel Design, Freeport, Illinois, MICRO SWITCH, a Honeywell Division, 1984.

112. R. N. Parrish, Development of Design Guidelines and Criteria for User/Operator Transactions With Battlefield Automated Systems, Synetics, Fairfax, Virginia, August 1980.

113. H. L. Parris and V. T. McConville, Anthropometric Data Base for Power Plant Design, NP-1918-SR, Electric Power Research Institute, Palo Alto, California, 1981.

114. R. N. Parrish et al., "Development of Design Guidelines and Criteria for User/Operator Transactions With Battlefield Automated Systems," Provisional Guidelines and Criteria for the Design of User/Operator Transactions, 4 (Draft Final Report, Phase I), Alexandria, Virginia: U.S. Army Research Institute, 1981.

115. S. D. Parsons, S. K. Ekert, and J. L. Seminara, "Human Factors Design Practices for Nuclear Power Plant Control Rooms," Proceedings of the Human Factors Society Twenty-Second Annual Meeting, 1978.

116. R. W. Pew, A. W. Rollins, and G. A. Williams, "Generic Man-Computer Dialogue Specification: An Alternative to Dialogue Specialists," Proceedings of the International Ergonomics Association 1976, London: Taylor and Francis, pp. 251-254.

117. I. Polack, "Auditory Displays and Noise," R. W. Pew and P. Green (eds.) Human Factors Engineering Short Course Notes, Twenty-Fifth Edition, Ann Arbor, Michigan: University of Michigan, Chrysler Center for Continuing Engineering Education, 1984.

118. K. M. Potosnak. Human Factors, "10 Tips for Getting Useful Information from Users," IEEE Software, pp. 89-90, July 1988.

119. Proceedings of the International Workshop: Ergonomic Aspects of Visual Display Terminals, Milan, March 1980, London: Taylor & Francis Ltd., 1983.

120. H. R. Ramsey and M. E. Atwood, Human Factors in Computer Systems: A Review of the Literature, Technical Report SAI 79-111-DEN, Science Applications, Inc., McLean, Virginia, 1979.

121. V. M. Reading (ed.), Visual Aspects and Ergonomics of Visual Display Units, University of London, Department of Visual Science, Institute of Opthalmology, 1978.

122. S. Rehulund, Ergonomic (translated from the Swedish by C. Soderstrom), Bildungskoncern: A. B. Volvo, 1973.

123. R. E. Richards and L. N. Haney, FAST: Operator—DPSE Interface Notes, unpublished Technical Report (Available from Robert_E. Richards, Idaho National Engineering Laboratory, EG&G Idaho, Inc., Idaho Falls, Idaho), U.S. Department of Energy, October 1986.

124. M. W. Riley, D. J. Cochran, and J. L. Ballard, "An Investigation of Preferred Shapes for Warning Labels," Human Factors, 24 (6), 1982, pp. 737-742.

125. W. Schiff, Perception: An Applied Approach, Boston: Houghton Mifflin Co., 1980.

126. H. G. Schultz, "An Evaluation of Format for Graphic Trend Displays—Experiment II," Human Factors, 3, 1961, pp. 99-107.

127. D. L. Schurman, "Panel Discussion Comment," Nuclear News, 31 (10), 1988 Annual ANS Meeting, August 1988.

128. R. Sekuler and P. D. Tynan, Sourcebook of Temporal Factors Affecting Information Transfer from Visual Displays, Technical Report 540, U.S. Army Research Institute for the Behavioral and Social Sciences, Alexandria, Virginia, June 1981.

129. C. A. Semple et al., Analysis of Human Factors Data for Electronic Flight Display Systems, AS-884-770, Manned Systems Sciences, Inc., Northridge, California, January 1971.

130. H. Shahnavaz, "Lighting Conditions and Workplace Dimensions of VDU Operators," Ergonomics, 25, 12, 1982, pp. 1165-1173.

131. R. S. Shirley, private communication, The Foxboro Co., Foxboro, Massachusetts, October 15, 1983.

132. D. A. Shurtleff, How to Make Displays Legible, La Mirada, California: Human Interface Design, 1980.

133. R. S. Sleight, "The Relative Discriminability of Several Geometric Forms," Journal of Experimental Psychology, 43, 1952, p. 424.

134. S. L. Smith, "Standards Versus Guidelines for Designing User Interface Software," Behavior and Information Technology, 5 (1), 1986, pp. 47-61.

135. S. L. Smith and J. N. Mosier, Guidelines for Designing User Interface Software, ESD-TR-86-278, Hanscom Air Force Base, Massachusetts: Electronic Systems Division, AFSC, United States Air Force, August 1986.

136. S. L. Smith and N. C. Goodwin, "Blink Coding for Information Display," Human Factors, 13 (3), 1971, pp. 283-290.

137. K. U. Smith and W. M. Smith, Perception and Motion: An Analysis of Space Structured Behavior in Gregory's R. L. Eye and Brain, New York: McGraw-Hill Book Co., 1966.

138. S. L. Smith and D. W. Thomas, "Color Versus Shape Coding in Information Displays," Journal of Applied Psychology, 48, 1964, p. 137.

139. H. L. Snyder and M. E. Maddox, Information Transfer from Computer-Generated Dot-Matrix Displays, HFL-78-3/ARO-78-1, Virginia Polytechnic Institute and State University, Blacksburg, Virginia, October 1978.

140. R. S. Soar, "Height-Width Proportions and Stroke Width in Numeral Visibility," The Journal of Applied Psychology, 39, 1955, pp. 43-46.

141. Software: Understanding Computers, Alexandria, Virginia: Time-Life Books, Inc., 1985.

142. R. W. Swezey and E. G. Davis, A Case Study of Human Factors Guidelines in Computer Graphics, IEEE CG&A, pp. 21-30, The Institue of Electrical and Electronic Engineers, Inc. New York, New York, November 1983.

143. "Invasion of the Data Snatchers—A Virus Epidemic Strikes Terror in the Computer World" Time, 123 (13), September 26, 1988.

144. M. A. Tinker, Legibility of Print, Ames, Iowa: Iowa State University Press, 1963.

145. J. R. Williams and R. P. Falzon, "Comparison of Search Time and Accuracy Among Selected Outlined Geometric Forms," Journal of Engineering Psychology, 2, July 1963, pp. 112-1128.

146. U.S. Department of Defense, Human Engineering Design Criteria for Military Systems, Equipment, and Facilities, MIL-STD-1472C, May 2, 1981.

147. U.S. Department of Defense, "Markings for Aircrew Station Displays," Design and Configuration, MIL-M-18012.

148. U.S. Department of Defense, Human Factors Engineering, AFSC-DH-1-3, Third Edition, Revision 1, Department of the Air Force, Wright-Patterson Air Force Base, Dayton, Ohio, June 1980.

149. U.S. Department of Defense, Human Engineering Guidelines for Management Information Systems, MIL-HDBK 761A (Draft), Proposed Revision, Aberdeen Proving Ground, Maryland: U.S. Army Human Engineering Laboratory, June 28, 1985.

150. U.S. Nuclear Regulatory Commission, Guidelines for Control Room Design Reviews, NUREG-0700, September 1981.

151. U.S. Nuclear Regulatory Commission, Human Factors Acceptance Criteria for the Safety Parameter Display System (Draft), NUREG-0835, October 1981.

152. U.S. Nuclear Regulatory Commission, Guidelines for the Preparation of Emergency Operating Procedures, NUREG-0899, August 1982.

153. H. P. Van Cott and R. G. Kinkade, Human Engineering Guide to Equipment Design, Washington, D.C.: U.S. Government Printing Office, 1972.

154. A. G. Vartabedian, "Effects of Parameters of Symbol Formation on Legibility," Information Display, 7 (5) May, 1970, pp. 23-26.

155. L. H. J. M. Verhagen, "Experiments with Bar Graph Process Supervision Displays on VDUs," Applied Ergonomics, 12,1, 1981, pp. 39-45.

156. D. L. Warren and M. J Burns, Lessons Learned From an Application of Programmable Display Push Buttons to Space Transportation System Procedures, JSC-22207, National Aeronautics and Space Administration, Lyndon B. Johnson Space Center, Houston, Texas June 1986.

157. Y. Waern and C. Rollenhagen, "Reading Test from Visual Display Units (VDUs)," Journal of Man Machine Studies, 18, 1983, pp. 441-465.

158. C. D. Wickens, Engineering Psychology and Human Performance, Toronto: Charles & Merril Publishing Company, 1984.

159. C. C. Woods, et al., Security Checklist for Computer-Based Information Systems for the Air Force Logistics Command, UCAR-10135, Air Force Logistics Command, Wright Patterson Air Force Base, Ohio, 1987.

160. W. E. Woodson, Human Factors Design Handbook, New York: McGraw-Hill, 1981.

ESSENTIAL USER-COMPUTER GUIDELINES

	Yes	No	N/A	Comments
VIDEO DISPLAYS				
HARDWARE ASPECTS				
Flicker				
1. The regeneration rate for a particular CRT display should be above the critical frequency of fusion so that the occurrence of disturbing flicker is not perceptible.				
Contrast Ratio				
1. 3:1 minimum; 5:1 to 10:1 preferred; 15:1 maximum.				
2. Background levels 15 to 20 cd/m^2.				
Display Luminance				
1. 45 cd/m^2 minimum, with 80 to 160 cd/m^2 preferred.				
2. 10 cd/m^2 minimum average level.				
Phosphor				
1. A green phosphor should be used.				
2. A medium persistence phosphor should be used.				
Glare				
1. The screen should be positioned so that sources of light and/or bright objects do not reflect into the expected viewing position.				

ESSENTIAL USER-COMPUTER GUIDELINES

	Yes	No	N/A	Comments
2. The surface of the VDU screen should be modified to reduce specular glare.				
Screen Resolution				
1. Regardless of whether the display is raster scanned or directly addressed, it should maintain the illusion of a continuous image; the viewer should not have to resolve scan lines or matrix spots.				
SCREEN STRUCTURES AND CONTENT				
Cursor				
1. The cursor should be easily seen but should not obscure the reading of the character or symbol it marks.				
2. The cursor should be easy to move from one position to another.				
3. The cursor should blink at about 3 Hz if it is used to attract the operator's attention on a monitoring task.				
4. The cursor should not be so distracting as to impair the searching of the display for information unrelated to the cursor.				
Text				
1. Consistent format should be maintained from one display to another.				
2. Prose should be displayed conventionally, in mixed upper and lower case.				

	Yes	No	N/A	Comments
3. Displayed paragraphs should be separated by at least one blank line.				
4. In textual display, every sentence should end with a period.				
5. In textual display, short, simple, concise sentences should be used.				
Labels				
1. Labels should convey the basic information needed for proper identification, utilization, actuation, or manipulation of the item.				
2. Labels should be consistent with such factors as a. Accuracy of identification required, b. Time available for recognition or other responses, c. Distance at which the labels must be read, d. Illuminance level and color, e. Criticality of the function labeled, and f. Vibration/motion environment of the user.				
3. Labels should be horizontal and read from left to right.				
4. Labels should be placed on or very near the items which they identify.				
5. For single data fields, the label should be placed to the left of the data field.				
6. For repeating data fields, the label should be placed above the data fields.				

ESSENTIAL USER-COMPUTER GUIDELINES

	Yes	No	N/A	Comments
7. Labels should primarily describe the functions of items.				
8. Control labeling should indicate the functional result of control movement (e.g., increase, ON, OFF).				
9. Control and display labels should convey verbal meaning in the most direct manner by using simple words and phrases. Abbreviations may be used when they are familiar to operators (e.g., psi, km).				
10. Words should be chosen on the basis of operator familiarity whenever possible, provided the words express exactly what is intended.				
11. Similar names for different controls and displays should be avoided.				
12. The units of measurement (e.g., volts, meters) should be labeled on the screen or panel.				
13. Labels should be printed in all capitals; periods should not be used after abbreviations.				
14. When dealing with mechanical labeling, to reduce confusion and operator search time, labels should be graduated in size.				
15. Label names should be easily discriminated from surrounding labeled fields or messages.				

ESSENTIAL USER-COMPUTER GUIDELINES

	Yes	No	N/A	Comments
16. Labels for data fields should be distinctively worded or highlighted so that they will not be readily confused with data entries, labeled control options, guidance messages, or other displayed material.				
17. Where entry fields are distributed across a display, a consistent format should be adopted for relating labels to entry areas.				
18. Where a dimensional unit (gpm, cm, deg) is consistently associated with a particular data field, it should be part of the fixed label, not entered by the user.				
Messages				
1. The computer should be capable of providing two levels of detail.				
2. Messages should be strictly factual and informative.				
3. Message dialogue should not be hostile to the user.				
4. Messages should be constructed using short, meaningful, and common words.				
5. The message should consider the prior knowledge of the user and the user's context.				
6. Sentences should be kept as simple in structure as possible.				

ESSENTIAL USER-COMPUTER GUIDELINES

	Yes	No	N/A	Comments
7. Messages should require no transformations, computing, interpolation, or reference searching.				
8. Messages should be stated in the affirmative and preferably in the active voice.				
9. Items to be remembered by the user should be placed at the beginning of the message.				
10. Items to be recalled by the user should be placed at the end of the message.				
11. Items of lesser importance should be placed in the middle of the message.				
Abbreviations				
1. Only standard and commonly accepted abbreviations should be used.				
2. Abbreviations should be short, meaningful, and distinct.				
3. The system should permit abbreviations of inputted commands.				
4. Whenever possible, experienced users should be provided with a set of abbreviations for frequently used commands.				
5. Abbreviations should be consistent in form.				
6. A dictionary of abbreviations should be available for on-line user reference.				
7. Abbreviations and acronyms should not include punctuation.				

ESSENTIAL USER-COMPUTER GUIDELINES

	Yes	No	N/A	Comments
Error Statements				
1. The computer system should contain prompting and structuring features by which an operator can request corrected information when an error is detected.				
2. Error messages should be worded as specifically as possible. When the computer detects an entry error, an error message should be displayed to the user stating what is wrong and what can be done about it; e.g., "No left parenthesis used."				
3. The wording of error messages should be appropriate to a user's task and level of knowledge.				
4. When a data entry or (more often) a control entry must be made from a small set of alternatives, those correct alternatives should be indicated in the error message displayed in response to a wrong entry.				
5. Error messages should be stated in polite but neutral wording, without implications of blame to the user, without personalization of the computer, and without attempts at humor.				
6. Following the output of simple error messages, the user should have the option of requesting more detailed explanation for errors (i.e., successively deeper levels of explanation				

	Yes	No	N/A	Comments
should be provided in response to repeated user requests for HELP).				
7. When multiple errors are detected in a combined user entry, some indication should be given to the user, even though complete messages for all errors cannot be displayed together.				
8. Error messages should be output 2 s after a user's entry has been completed.				
9. System documentation should include, as a supplement to on-line guidance, a listing and explanation of all error messages.				
10. Following error detection, users should be prompted to reenter only the portion of a data/command entry that is not correct.				
11. In addition to a clear text error message, an error identification number (ID) should precede each message.				
12. Error messages should always state or clearly imply at least a minimum of (a) what error has been detected and (b) what corrective action to take.				
13. If an error is detected in a group of stacked entries, the system should process correct commands until the error is displayed. A suitable error message should be presented, and no more inputs should be processed until the error is corrected.				

ESSENTIAL USER-COMPUTER GUIDELINES

Nontextual Messages	Yes	No	N/A	Comments
1. When using alphanumeric codes, a consistent convention should be adopted that all letters shall be either uppercase or lowercase.				
2. When codes combine letters and numbers, characters of each type should be grouped together rather than interspersed.				
3. Meaningful codes should be adopted in preference to arbitrary codes; e.g., a three-letter mnemonic code (DIR = directory) is easier to remember than a three-digit numeric code.				
4. When arbitrary codes must be remembered by the user, they should be no longer than four to five characters.				
5. Code length and format should be constant throughout any single category.				
6. Codes should contain predictable letter sequences.				
7. Long codes (seven or more characters) should be broken into three- or four-character groups; i.e., separate groups by a hyphen or blank space.				
8. Refrain from using 1's and 0's in code vocabularies.				

ESSENTIAL USER-COMPUTER GUIDELINES

Data Display

	Yes	No	N/A	Comments
1. Displayed data should be tailored to user needs, providing only necessary and immediately usable information at any step in a transaction sequence.				
2. Data should be displayed to the user in directly usable form.				
3. Data should be consistent, following standards and conventions familiar to the user.				
4. When protection of displayed data is essential, do **not** permit a user to change controlled items.				
5. In general, do not require the user to rely on memory, but recapitulate needed items on the succeeding display.				
6. The detailed internal format of frequently used data fields should be consistent from one display to another.				
7. Long data items of arbitrary alphanumeric characters should be displayed in groups of three or four separated by a blank.				
8. In tabular displays, columns and rows should be labeled following the same guidelines proposed for labeling the fields of data forms.				
9. In tabular displays, the units of displayed data should be consistently included in the column labels or following the first row of entry.				

ESSENTIAL USER-COMPUTER GUIDELINES

	Yes	No	N/A	Comments
10. Columns of numeric data without decimals should be displayed right-justified; numeric data with decimals should be justified with respect to the decimal point.				
11. Lists of alphabetic data should be vertically aligned with left- justification to permit rapid scanning; indentation can be used to indicate subordinate elements in hierarchic lists.				
12. Data lists should be organized in some recognizable order, whenever feasible, to facilitate scanning and assimilation; e.g., dates may be ordered chronologically, names alphabetically.				
13. Listed data should be distinctive from lists of menu options.				
14. When listed items are labeled by number, the numbering should start with 1 and not 0.				
15. For hierarchic lists with compound numbers, the complete numbers should be used, rather than omitting the repeated elements; i.e., 2.1 Position Designation 2.1.1 Arbitrary Positions 2.1.1.1 Discrete 2.1.1.2 Continuous				
16. In dense tables with many rows, a blank line (or some other distinctive feature) should be inserted after every fifth row as an aid for horizontal scanning. (If space permits, a blank line after every third row is even better).				

ESSENTIAL USER-COMPUTER GUIDELINES

	Yes	No	N/A	Comments
17. When data are displayed in more than one column, the columns should be separated by at least three to four spaces if right-justified and by at least five spaces otherwise.				
18. When tables are used for referencing purposes, such as an index, the indexed material should be displayed in the left column, the material most relevant for user response in the next adjacent column, and associated but less significant material in columns further to the right.				
19. Longer series of strings or lists of data should be organized in columns to provide better legibility and faster scanning.				
20. If data are to be entered from paper forms, the design of the input screen and the layout of the paper form should correspond. This helps the user to find and keep a location while looking back and forth from the form to the terminal.				
21. Each list of selections should have a heading that reflects the question for which an answer is sought; e.g., Select Plant Mode 1. Start up 2. Steady-state operation 3. Shutdown				
22. In a list of options, the most frequently used options should be placed at the top of the list.				

ESSENTIAL USER-COMPUTER GUIDELINES

	Yes	No	N/A	Comments
23. Selection numbers should be separated from text descriptors by at least one space. Include space after the period, if used. Right-justify selection numbers.				
24. When lists or data tables extend beyond one display page, the user should be informed when a list is or is not complete.				
25. Labels for single data fields should be located to the left of the data field and separated from the data field by a unique symbol (such as a colon) and at least one space, e.g., TITLE: ___________				
26. When caption sizes are relatively equal, both captions and data fields should be justified left. One space should be left between the longest caption and the data field column; e.g., FEED FLOW: _________ PRIMARY COOLANT: ______				
27. When caption sizes vary greatly, captions should be right-justified and the data fields should be left-justified. One space should be left between each caption and the data field; e.g., FEED FLOW: ___ STEAM GENERATOR PRESSURE: ___				

	Yes	No	N/A	Comments
28. A feld group heading should be centered above the captions to which it applies. It should be completely spelled out and related to the captions; e.g., - - - - - DRYWELL - - - - - - LEVEL (FT.) ___ PRESSURE (PSIG) ___ TEMPERATURE (F) ___				
29. When section headings are located on the line above related screen fields, the captions should be indented a minimum of five spaces from the start of the heading; e.g., SECONDARY CONTAINMENT LEVEL: _____ PRESS: _____				
30. When section headings are placed adjacent to the related fields, they should be located to the left of the top most row of related fields. The column of captions should be separated from the longest heading by a minimum of three blank spaces; e.g., SECONDARY CONTAINMENT LEVEL: ____ PRESS: ____				
31. At least five spaces should appear between the longest data field in one column and the right most caption in an adjacent column; e.g., ALARM: ______ DATE: ______ STATUS: _____ TIME: ______				

	Yes	No	N/A	Comments
32. Where space constraints exist, vertical lines may be substituted for spaces for separation of columns of fields.				
33. For multiple-occurrence fields without group headings, at least three spaces should exist between the columns of fields; e.g., PWR LVL TEMP ____ ____ ____ ____ ____ ____ ____ ____ ____				
34. For multiple-occurrence fields with group headings, at least three spaces should appear between columns of related fields and at least five spaces should appear between groupings; e.g., —— UNIT I —— —— UNIT II —— PWR LVL TEMP PWR LVL TEMP ___ ___ ___ ___ ___ ___ ___ ___ ___ ___ ___ ___ ___ ___ ___ ___ ___ ___				
Data Entry				
1. When form filling, the user should be allowed to RESTART, CANCEL, or BACKUP and change any item before taking a final ENTER action.				
2. Whenever possible, multiple data items should be entered without the need for special separators or delimiters, either by keying into predefined entry fields or by including simple spaces between sequentially keyed items.				

	Yes	No	N/A	Comments
3. When a field delimiter **must** be used for data entry, a standard character should be adopted for that purpose; a slash (/) is recommended.				
4. For all dialogue types involving prompting, data entries should be prompted explicitly by displayed labels for data fields and/or by associated user guidance messages; e.g., NAME: _ _ _ _ _ _ _ _ _ _ _ ORGANIZATION: _ _ / _ _ PHONE: _ _ _ / _ _ _ _				
Data Field Guidelines				
5. Field labels should consistently indicate what data items are to be entered.				
6. In ordinary use, field labels should be protected and transparent to keyboard control so that the cursor skips over them when spacing or tabbing.				
7. Special characters should be used to delineate each data field; a broken-line underscore is recommended; e.g., Enter plant code: _ _ _ _				
8. Implicit prompting by field delineation should indicate a fixed or maximum acceptable length of the entry; e.g., Enter ID : _ _ _ _ _ _ _ _				
9. Input prompts should indicate which entries are mandatory and which are				

	Yes	No	N/A	Comments
optional. Mandatory fields should be located ahead of optional ones. All inputs should be mandatory unless they are marked optional. Enter plant code: ________ Region (optional): __________ Architectural Engineer (optional): ___				
10. When item length is variable, the user should not have to justify an entry either right or left and should not have to remove any unused underscores; computer processing should handle those details automatically.				
11. When multiple items (especially those of variable length) will be entered by a skilled touch typist, each data field should end with an extra (blank) character space; software should be designed to prevent keying into a blank space, and an auditory signal should be provided to alert the user when that happens.				
12. Labels for data fields should be distinctively worded so that they will not be readily confused with data entries, labeled control options, guidance messages, or other displayed material.				
13. When displayed data forms are crowded, auxiliary coding should be adopted to distinguish labels from data.				
14. In labeling data fields, only agreed terms, codes, and/or abbreviations should be used.				

	Yes	No	N/A	Comments
15. The label for each entry field should end with a special symbol, signifying that an entry may be made; A colon is recommended for this purpose, e.g.: NAME: _ _ _ _ _ _ _				
16. Labels for data fields may incorporate additional cueing of data formats when that seems helpful; e.g., DATE (M/D/Y): _ _ / _ _ / _ _ DATE:_ _ / _ _ / _ _ M M D D Y Y				
17. When a measurement unit is consistently associated with a particular data field, it should be displayed as part of the fixed label rather than entered by the user. (Data should be keyed without dimensional units, e.g., gpm, psig, Klb/hr, etc.) HIGH PRESSURE INJECTION (GPM): _ _ _ _ _ _ MAIN FEED CONTROL (PCT OPEN): _ _ _ _ _				
Data Familiarity and Sequence				
18. Data should be entered in units that are familiar to the user.				
19. When data entry involves transcription from source documents, form-filling displays should match (or be compatible with) paper forms; in a question-				

	Yes	No	N/A	Comments
and-answer dialogue, the sequence of entry should match the data sequence in source documents.				
20. If no source document or external information is involved, the ordering of multiple-item data entries should follow the logical sequence in which the user is expected to think of them.				
21. When a form for data entry is displayed, the cursor should be positioned automatically in the first entry field.				
22. When sets of data items must be entered sequentially in a repetitive series, a tabular format where data sets are keyed row by row should be used.				
23. Justification of tabular data entries should be handled automatically by the computer; the user should **not** have to enter any leading blanks or other formatting characters; e.g., if a user enters 56 in a field four characters long, the system should **not** interpret 56 as 5600.				
24. It should be possible for the user to make numeric entries (e.g., dollars and cents) as left-justified, but they should be automatically justified with respect to a fixed decimal point when a display of the data is subsequently regenerated for review by the user.				
25. For dense tables (those with many row entries), some extra visual cue should be provided to guide the user accurately across columns.				

	Yes	No	N/A	Comments
26. Software for automatic data validation should be incorporated to check any item whose entry and/or correct format or content is required for subsequent data processing.				
<u>Transactions, Cross Referencing, and Flexibility</u>				
27. In a repetitive data entry task, data validation for one transaction should be completed and the user allowed to correct errors before another transaction begins.				
28. When helpful values for data entry cannot be predicted by user system interface (USI) designers, which is often the case, the user (or perhaps some authorized supervisor) should have a special transaction to define, change, or remove default values for any data entry field.				
29. On initiation of a data entry transaction, currently defined default values should be displayed automatically in their appropriate data fields.				
30. User acceptance of a displayed default value for entry should be accomplished by simple means, such as by a single confirming key action or simply by tabbing past the default field.				
31. A user should not be required to enter bookkeeping data that the computer could determine automatically. For example, a user generally should not have to identify his workstation to				

	Yes	No	N/A	Comments
initiate a transaction, nor include other routine data such as transaction sequence codes.				
32. A user should not be required to enter redundant data already accessible to the computer. The user should **not** have to enter such data again. For example, the user should not have to enter both an item name and identification code when either one defines the other.				
33. Whenever needed, automatic cross-file updating should be provided so that a user does not have to enter the same data twice.				
34. When data entry requirements change, which is often the case, some means should be provided for the user (or an authorized supervisor) to make necessary changes to data entry procedures, entry formats, data validation logic, and other associated data processing.				
35. Areas of the screen not containing entry fields (i.e., protected fields) should be inaccessible to the operators and not require repeated key depressions to step through.				
36. Space lines should be incorporated where visual breaks or spaces occur on the source document.				
37. A section heading should be located directly above its associated fields.				

ESSENTIAL USER-COMPUTER GUIDELINES

	Yes	No	N/A	Comments
38. When possible, stacking of input or multiple entries should be permitted; e.g.,				
39. The user should be able to alter input during and after entry.				
40. In a variable-length entry, the user should be required to enter only the relevant input data.				
41. When possible, a system should recognize common misspellings of a command and execute the command as if it had been spelled correctly.				
42. Misspelling of similar commands should not cause errors.				
43. Keying should be minimized.				
44. The user should not be required to reenter parameters that have not changed since the previous interaction.				
Instructions—General				
1. Words in instructions should be meaningful to the user.				

Item 38 example:

Code	Category
LXX	Labor
MXX	Material

Selection Code: LXX; LRX; LRT (The last two codes normally are displayed at the next two menu levels, but three entries are stacked here to save time.)

ESSENTIAL USER-COMPUTER GUIDELINES

	Yes	No	N/A	Comments
2. Short words should be used in instructions.				
3. Active voice and the affirmative case should be used in instructions.				
4. Instructions should be patterned.				
Instructions—Illustrations				
1. Illustrations should be appropriate for the type of information to be conveyed.				
2. Illustrations should be placed close to the corresponding text.				
3. Wording on illustrations should be minimized.				
4. Tables and graphs should be captioned.				
Instructions—Style and Structure				
1. When instructions must be rapidly accessed, a table of contents and/or an index should be provided.				
2. The literary style of a set of instructions should be appropriate to its intended use.				
3. Instructions should have a clearly stated beginning and a well-developed summary.				
4. Paragraphs of text should be short and should contain a single idea.				
5. Instructions should be simple.				

ESSENTIAL USER-COMPUTER GUIDELINES

	Yes	No	N/A	Comments
6. Instructions should state important items more than once.				
7. Instructions should contain only essential information.				
8. The amount of detail should be appropriate to the experience of the user.				
9. The sequence of the instructions should follow the sequence of actions required.				
10. Short sentences, flow diagrams, algorithms, lists, and tables are superior to prose.				
11. The main topic of the instruction should appear at the beginning of the sentence.				
12. All instructions should be tested on naive users before being finalized.				
13. Many-step instructions should use a two-column format.				
14. In a list of specifications for service or supply, more than a part number should be given.				
15. Warning and caution notices should be accurate and concise and should contain only the information relevant to the warning or caution. They should not contain operator actions.				
16. Warnings and cautions should immediately precede the steps to which they refer.				

	Yes	No	N/A	Comments

CHARACTERISTICS OF ALPHANUMERIC CHARACTERS

Font or Style

1. In designing a visual display character set, each character should be designed so that fine differences in stroke length, curvature, etc., are preserved in order to avoid similarity.

2. For a given font, it should be possible to clearly distinguish between the following characters:

   X and K I and L
   O and Q U and V
   T and Y I and 1
   S and 5

Character Size and Proportion

1. Character height should be 16 min of arc to 26.8 min of arc, with 20 min of arc preferred.

2. The ratio between character height and width should be from 1:1 to 5:3.

3. The ratio between character height and stroke width should be from 5:1 to 8:1.

4. The minimum spacing between characters should be one stroke width.

5. The minimum spacing between words should be one character width.

6. Spacing between lines should be from 50 to 150% of the character height.

ESSENTIAL USER-COMPUTER GUIDELINES

	Yes	No	N/A	Comments
Character Case				
1. Labels or statements should be in upper case.				
2. Text should be displayed in both uppercase and lowercase.				
SCREEN ORGANIZATION AND LAYOUT				
Screen Size				
1. The screen should be the smallest size which will allow required information to be seen clearly and easily by the viewer.				
2. The screen should take into account the distance of the operator from the screen (e.g., large screen overviews).				
Grouping				
1. Information that is continually being transmitted or received should be **sequentially** grouped.				
2. Information should be grouped in the order of its **frequency** of use.				
3. If frequency of use is not a major concern, information should be **functionally** grouped.				
4. When some items are more critical than others to the success of the systems, the information should be grouped by **importance**.				
5. Grouped data should be arranged in				

	Yes	No	N/A	Comments
the display with consistent placement of items so that user detection of similarities, differences, trends, and relationships is facilitated.				
6. When there is no appropriate logic for grouping data (sequence, function, frequency, or importance), some other principle should be adopted, such as alphabetical or chronological grouping.				
7. Similar information should be displayed in groups according to the left-to-right or top-to-bottom rules.				
8. All displayed data necessary to support an operator activity or sequence of activities should be grouped together.				
9. Screens should provide cohesive groupings of screen elements so that people perceive large screens as consisting of smaller identifiable pieces. People are relatively efficient at viewing groups or chunks of data.				
10. Providing perceptual structure on the screen can be achieved in a variety of ways, ranging from some arbitrary but consistent grouping to an optimally designed functional grouping based upon fequency of usage data.				
11. Grouping similar items together in a display format improves their readability and can highlight relationships between different groups of data. Grouping can be used to provide structure in the display and aid in the recognition and identification of specific items of information.				

ESSENTIAL USER-COMPUTER GUIDELINES

	Yes	No	N/A	Comments
Display Density				
1. Screen packing density should not exceed 50% and preferably should be less than 25%.				
2. Display screens should be perceived as uncluttered.				
3. Provide information that is only essential to making a decision or performing an action.				
4. All data related to one task should be placed on a single screen.				
5. For critical task sequences, screen packing density should be minimized.				
6. Where user information requirements cannot be accurately determined in advance of interface design or are variable, on-line user options should be provided for data selection, display coverage, and suppression.				
Display Partitioning/Windows				
1. Screens should be divided into windows that are clearly perceptible to the user.				
2. On large, uncluttered screens, windows should be separated using three to five rows or columns of blank space.				
3. Specific areas of the screen should be reserved for information such as commands, status messages, and input fields; those areas should be consistent on all screens.				

	Yes	No	N/A	Comments
4. When a display window must be used for data scanning, the window size should be greater than one line.				
5. The screen should not be divided into a large number of small windows.				
6. When the body of the display is used for data output, the screen should be coherently formatted and not partitioned into several small windows.				
7. The number of overlapping windows should be minimized.				
8. If possible, program windows so that the size is expandable by the user.				
Frame Specifications				
1. Specific areas of the screen should be reserved for information such as commands, status messages, and input fields; those areas should be consistent on all screens.				
2. Both the items on display and the displays themselves should be standardized.				
3. An invariant field, including the page title, an alphanumeric designator, the time, and the date, should be placed at the top of each display page.				
4. The last four lines (at least) of each display page should be reserved for variant fields.				

	Yes	No	N/A	Comments
5. Procedures for user actions should be standardized.				
6. Each display frame should have a unique identification (ID).				
7. Every frame should have a title on a line by itself.				
8. Status information should be displayed near the top-right corner of the screen.				
9. Location coding should be employed to reduce operator information search time.				
Interframe Considerations —Paging and Scrolling				
1. Whenever possible, all data relevant to the user's current transaction should be included in one display page (or frame).				
2. When the requested data exceed the capacity of a single display frame, the user should be provided easy means to move back and forth among relevant displays either by paging or scrolling.				
3. When a list of numbered items exceeds one display page and must be paged/scrolled for its continuation, items should be numbered continuously in relation to the first item in the first display and should indicate the present maximum location. For example, use Line 63 of 157, not Page 3, Line 8.				

	Yes	No	N/A	Comments
4. When lists or tables are of variable length and may extend beyond the limits of a single display page, their continuation and ending should be explicitly noted on the display. For example, incomplete lists might be marked "continued on next page," or simply "continued." Concluding lists might add a note "end of list."				
5. When display output contains more than one page, the notation "page x of y" should appear on each display.				
6. The parameters of roll/scroll functions should refer to the data being reviewed, not to the window. For example, "roll up 5 lines" should mean that the top five lines of data would disappear and five new lines would appear at the bottom; the window through which the data is viewed remains fixed.				
7. When the user may be exposed to different systems adopting different usage, any reference to scroll functions should consistently use functional terms such as forward and back (or next and previous) to refer to movement within a displayed data set rather than words implying spatial orientation (e.g., up and down).				
8. When using a menu system, the user at all times should have access to the main menu.				
9. Displays should indicate how to continue.				

ESSENTIAL USER-COMPUTER GUIDELINES

	Yes	No	N/A	Comments
10. User-terminal interaction tasks that are repetitive, time-consuming, or complex should be assigned dedicated functions.				
11. Required or frequently used data elements should be included on the earliest screens in the application transaction.				
12. Page design and content planning should minimize requirements for operator memory.				
<u>Frame Hierarchy and Elements</u>				
13. When pages are organized in a hierarchical fashion containing a number of different paths through the series, a visual audit trail of the choices should be available upon operator request.				
14. Sectional coordinates should be used when large schematics must be panned or magnified.				
15. If the message is a variable option list, common elements should maintain their physical relationship to other recurring elements.				
16. A message should be available that provides explicit information to the user on how to move from one frame to another or how to select a different display.				
17. When the operator must step through multiple display levels, priority access should be provided to the more critical display levels.				

ESSENTIAL USER-COMPUTER GUIDELINES

	Yes	No	N/A	Comments
18. When the operator must step through multiple display levels, he or she should be provided with information identifying the current position within the sequence of levels.				
19. A similar display format should be used at each level of a multiple-level display.				
20. When the operator is required to accurately comprehend previously learned items appearing with a new list, the list should be kept small (about four to six items).				
21. Frequently appearing/disappearing commands/subcommands should be placed in the same place on the screen.				
Interframe Considerations —Windowing				
1. The system should not allow for more than three applications to be run at a time.				
2. Once you quit a program (application), that window should close promptly.				
3. Windows should be consistent in their use of drop-down menus and/or icons.				
4. Navigation within the menus, whether mouse or direction keys are used, must be consistent.				
5. Windows should give the user feedback whenever he/she is in the process of combining applications.				

	Yes	No	N/A	Comments
6. Dialogue boxes should be provided when necessary, to assist in defining menu options.				
7. Actions necessary for changing the size of a window should be consistent between windows.				
8. Active windows should be so labeled.				
9. Window labels should be located at the top of the window border.				
10. Keyboard input should only affect the **active** window.				
VISUAL CODING DIMENSIONS				
Color				
1. Color should be used as a formatting aid to assist in structuring a screen and as a code to categorize information or data.				
2. Color coding should not create unplanned or obvious new patterns on the screen.				
3. Color coding should be applied as an additional aid to the user on displays that have already been formatted as effectively as possible on a single color.				
4. When color coding is used, it should be redundant with some other feature in data display, such as symbology.				

	Yes	No	N/A	Comments
<u>User Control</u>				
5. The unit should, as a minimum, be provided with the following controls:				
a. A foreground intensity control separate from the background intensity control.				
b. A capability for making grid lines half as intense as the rest of the display.				
c. Enough intensity control variable to accommodate very low ambient illumination and the higher levels (5 to 150 fc).				
<u>Color Consistency</u>				
6. Color meanings should be consistent with traditional color expectancies.				
7. Color coding should be consistent within a frame, from frame to frame, and with other color-coded systems in the control room (see Table 4).				
8. Color codes should conform to color meanings that already exist in the user's job.				
9. Specific color selections should conform to the general guidelines of Table 5 and to the following specific recommendations:				

ESSENTIAL USER-COMPUTER GUIDELINES

	Yes	No	N/A	Comments
a. The most generally used colors should be red, green, yellow, and blue. Other acceptable colors are orange, yellow-green, blue-green, and violet.				
b. When blue headings, numbers, or alphabetic characters are used, the background should be black.				
c. Yellow should not be used on a white background because of the very low contrast.				
d. Yellow should not be used on a green background due to a vibrating effect to the eye.				
e. White should be used for very important information.				
f. The selected colors should yield satisfactory color contrast for color-deficient users.				
g. The user should be able to discriminate the selected color on an absolute basis.				
h. Selected colors should be usable in all control room applications (e.g., panel surfaces, labels, CRTs, indicator light bulbs or filters, console surfaces).				
i. Blue should be used only for background features in a display, **not** for critical data.				
j. Whenever possible, red and green should not bewused in combina-				

	Yes	No	N/A	Comments
tion. Usewof red symbols/characters on a green background especially should be avoided.				
10. If a pattern of color is intended to display a function, the selected colors should indicate the state of the system.				
11. Colors with high contrast should be selected for parameters and features that must "catch" the operator's attention.				
12. In general, backgrounds should not be brighter than foregrounds.				
13. Extreme color contrasts should be avoided.				
14. Colors should be specified as a precise wavelength rather than a hue (red, green, violet, etc.) (see Table 6).				
15. If difference in brightness (intensity) is used as a coding mechanism, perceived brightness should be used rather than absolute brightness.				
16. Each color should represent only **one** category of displayed data.				
<u>Number of Colors</u>				
17. If color discrimination is required, do not use more than eight colors. Alphanumeric screens should display no more than four colors at one time.				

	Yes	No	N/A	Comments
Control Room Lighting				
18. Colored ambient lighting should not be used in conjunction with color-coded CRTs.				
19. If ambient illumination cannot be controlled, hoods should be provided that block out light and glare.				
20. High-pressure sodium should not be used as an ambient-light medium for CRT viewing.				
Maintenance of Colors				
21. Color displays should be periodically adjusted to maintain proper registration of images.				
Color Applications for Mimics				
22. Identify potential applications of color in mimic designs. Mimics can incorporate color to differentiate process flow paths. For example, **blue** may be used to code water lines; **white**, steam lines; **yellow**, oil lines; and so forth. Such color differential is potentially valuable in helping operators to sort out complex interrelationships.				
Color for Highlighting				
23. Color is extremely effective for highlighting related data which are spread around on a display, such as data of a particular status or category. Color may effectively aid in the location of headings, out-of-tolerance data, newly				

	Yes	No	N/A	Comments
entered data, data requiring attention, etc. Search for data on a display is facilitated by color coding if the color of the data sought is known.				
24. A monochromatic format can be used if the flash is twice as intense as the rest of the display. This will draw the operator's attention as effectively as if color had been added. Fewer colors can be used if they can be adjusted for intensity.				
Misuse of Color				
25. Do not overuse color. Use of too many colors may make a screen confusing or unpleasant to look at.				
26. Color capability must be used conservatively in the design of display screens. Arbitrary use of multiple colors may cause the screen to appear busy or cluttered, and it may reduce the likelihood that the information in color codes on that screen or on other screens will be interpreted appropriately and quickly. In general, color should be added to the base color display only if it will help the user in performing a task.				
27. Do **not** use color coding in an attempt to compensate for poor display format; redesign the display instead.				
28. The use of color should not be distracting to the user.				

ESSENTIAL USER-COMPUTER GUIDELINES

	Yes	No	N/A	Comments
Geometric Shape				
1. Geometric shapes should be considered for discriminating different categories of data on graphic displays.				
2. Alphabets of geometric shapes should be limited to a maximum of 15 different symbols.				
<u>Design of Shapes</u>				
3. When geometric shape (symbol) coding is used, the basic symbols should vary widely in shape.				
4. Symbols should subtend a minimum of 20 min of arc. If the viewing distance is longer than the normal 28 in., it should form a visual angle of about 22 min of arc.				
5. The stroke width-to-height ratio should be 1:8 or 1:10 for symbols of 0.4 in. or larger viewed up to a distance of 7 ft.				
<u>Redundancy</u>				
6. When efficiency of decoding is important, redundant cues (such as size difference) should be used.				
7. When rate of comprehension and detection is important, graphical coding should be used rather than word messages.				

ESSENTIAL USER-COMPUTER GUIDELINES

	Yes	No	N/A	Comments
<u>Standardization</u>				
8. The assignment of shape codes should be consistent for all displays and should be based upon an established standard.				
Pictorial				
1. Pictographs should have obvious meanings, and the meaning should be tested in the user population.				
2. Symbols should be consistently applied.				
<u>Mixing Pictographs and Words</u>				
3. Words and symbols should not be used alternately.				
4. Symbols should be used to represent equipment components and process flow or signal paths, along with numerical or coded data reflecting inputs and outputs associated with equipment.				
<u>Design of Pictographs</u>				
5. Symbols used to represent equipment components vary widely in shape and should be similar to those used in piping and instrumentation drawings.				
6. About six different symbols, with 20 being an upper limit, should be used.				
7. When iconic symbols are used, solid forms without unnecessary detail are preferred.				

ESSENTIAL USER-COMPUTER GUIDELINES

	Yes	No	N/A	Comments
8. The visual saliency of those features that must remain redundant across members of a symbol set should be minimized.				
9. A closed figure enhances the perceptual process and should be used unless there is reason for the outline to be discontinuous.				
Labeling Conventions				
10. When letters are used, perhaps to annotate geometric display symbols, lower-case letters should be used to improve discriminability.				
11. For search and identification tasks or whenever there is any doubt as to whether some observers will be able to understand the pictorial, both pictorial and word labels should be used.				
Redundancy				
12. Given that a sufficient number of dimensions are available to portray required information parameters, multidimensional display codes should be used.				
Distortion				
13. To minimize distortion (especially under degraded viewing conditions), well-learned or unitized symbol designs should be used.				

	Yes	No	N/A	Comments
Pictorial Simplicity				
14. The symbols should be as simple as possible, consistent with the inclusion of features that are necessary.				
Control Room Lighting				
15. The pictorial pattern should be identifiable from the maximum viewing distance and/or under minimal ambient lighting conditions.				
Orientation				
16. Pictorial symbols should always be oriented "upright."				
Iconic Representation				
17. Icons should not be used when display resolution is low.				
18. A label should be associated with each icon.				
19. Abstracts (icons) may be of either a literal, functional, or operations type.				
20. To the extent possible, icons should concur with existing population or industry stereotypes.				
Magnitude				
1. When symbol size is used for coding, the intermediate symbols should be spaced logarithmically between the two extremes (largest and smallest).				

	Yes	No	N/A	Comments
2. When the symbol size is to be proportional to the data value, the scaled parameter should be the symbol area rather than a linear dimension such as diameter.				
3. For AREA coding, the maximum number of code steps should be six, with three recommended.				
4. For LENGTH coding, the maximum number of code steps should be six, with three recommended.				
5. Size coding should be used only when displays are not crowded.				
6. When size coding is used, a larger symbol should be at least 1.5 times the height of the next smaller symbol.				
ENHANCEMENT CODING DIMENSIONS				
Brightness				
1. No more than three levels of brightness coding should be used, with two levels preferred. For example, a data form might combine bright data items with dim labels to facilitate display scanning.				
2. Brightness coding should be employed only to differentiate between an item of information and adjacent information.				
3. High brightness levels should be used to signify information of primary importance and lower levels to signify information of secondary interest.				

	Yes	No	N/A	Comments
4. Brightness coding should not be used in conjunction with shape or size coding.				
5. When an operation is to be performed on a single item on a display, the item should be highlighted.				
6. In a list, the option(s) selected by the user should be highlighted.				
7. Maximum contrast should be provided between those items highlighted and those not.				
8. When graphical items are close together on the screen, successive brightening of graphical items and user selection by button activation should be considered.				
Blink				
1. Blink coding should be limited to small fields.				
2. The blink rate should lie in the range of 0.1 to 5 Hz, with 2 to 3 Hz preferred.				
3. The minimum "on" time should be 50 ms.				
4. To avoid interference with reading performance, the blink rate should be such that the user can match the operator's scan rate to the blink rate.				
5. If difference in blink rate is used as a coding method, no more than two steps should be used.				

Guideline	Yes	No	N/A	Comments
6. When two blink rates are used, the fast blink rate should approximate four per second and the slow rate should be one blink per second.				
7. When two blink rates are used, the higher rate should apply to the most critical information.				
8. When two blink rates are used, the "on-off" ratio should approximate 50%.				
<u>Operator Control</u>				
9. A means should be provided for suppressing the blink action once the coded data have been located.				
10. An "off" condition should never be used to attract attention to a message.				
<u>Use of Blink Coding</u>				
11. Blinking should be reserved for emergency conditions or similar situations requiring immediate operator action.				
12. When blink coding is used to mark a data item that must be read, an extra symbol (such as an asterisk) should be added as a blinking marker rather than blinking the item itself.				
13. Blink coding should be used for target detection tasks, particularly with high-density displays.				
14. Blink coding should not be used with long-persistence phosphor displays.				

	Yes	No	N/A	Comments
Image Reversal				
1. Image reversal (e.g., dark characters on a light background) should be used primarily for highlighting in dense data fields.				
2. Image reversal can be used to code annunciator information that requires immediate response.				
3. Maximum contrast should be provided between highlighted and nonhighlighted items.				
Auditory				
1. Audio displays should be provided when				
a. The information to be processed is short, simple, and transitory, requiring immediate or time-based response.				
b. The common mode of visual display is restricted by overburdening; ambient light variability or limitation; operator mobility; degradation of vision by vibration, high g-forces, hypoxia, or other environmental considerations; or anticipated operator inattention.				
c. The criticality of transmission response makes supplementary or redundant transmission desirable.				
d. It is desirable to warn, alert, or cue the operator to subsequent additional response.				

	Yes	No	N/A	Comments
e. Custom or usage has created anticipation of an audio display.				
f. Voice communication is necessary or desirable.				
Audio System Reliability and Testing				
2. The design of audio display devices and circuits should preclude false alarms.				
3. The audio display device and circuit should be designed to preclude warning signal failure in the event of system or equipment failure and vice versa.				
4. Audio displays should be equipped with circuit test devices or other means of operability testing.				
Control Room Environment				
5. Audio displays should be compatible with the ambient conditions in which they are used.				
Criteria for Audio Signal Selection				
6. If a signal type is commonly associated with a certain type of activity, it should not be used for other purposes when the situation is such that the more common convention is in use.				
7. Once a particular auditory signal code is established for a given operating situation, the same signal should not be designated for some other display.				

	Yes	No	N/A	Comments
8. The frequency of an audio signal should be within the range of 200 to 5000 Hz, and preferably between 500 and 3000 Hz.				
9. When small changes in signal intensity must be detected, the signal frequency should be from 1000 to 4000 Hz.				
10. When an audio signal must travel over 1000 ft, its frequency should be less than 1000 Hz.				
11. When an audio signal must bend around major obstacles or pass through partitions, its frequency should be less than 500 Hz.				
12. When the noise environment is unknown or suspected of being difficult to penetrate, the audio signal should have a shifting frequency that passes through the entire noise spectrum and/or be combined with a visual signal.				
13. If a signal must occur in an area in which only certain personnel should be privy to its purpose and others are not to be unduly annoyed, a simple bell tone should be used that is recognizable among ambient speech sounds without being loud.				
14. When the signal must indicate which operator (of a group of operators) is to respond, a simple repetition code should be used.				
15. Audio signals should not startle listeners, add to overall noise levels, or interfere with local speech activity.				

ESSENTIAL USER-COMPUTER GUIDELINES

	Yes	No	N/A	Comments
16. Auditory signals should be easily discernible from any ongoing audio input (be it either meaningful input or noise).				
17. The intensity of audio signals should be at least 60 dB above the absolute threshold.				
18. When complex information is to be presented, two-stage signals should be used.				
19. If a person is to listen concurrently to two or more channels, the frequency of the channels should be different.				
20. Audio signals should provide only that information which is necessary for the user.				
21. Where feasible, interrupted or variable signals should be used rather than steady-state signals.				
22. Audio signals should be tested prior to using them.				
23. Auditory signal frequencies should differ from those that dominate any background noise.				
<u>Audio System Modification</u>				
24. When equipment is modified or added to an existing system, any new audio signals should be compatible with existing audio signals.				
25. When an audio signal is installed to replace another type of signal, a changeover period should be allowed				

	Yes	No	N/A	Comments
during which both the new and the old signal are in effect.				
Dedicated System for Warnings				
26. Where feasible, a separate communication system (such as loudspeakers, horns, or devices not used for other purposes) should be used for warnings.				
Voice				
1. A **tonal signal** should be used when				
a. Immediate action is required on the part of the listeners (i.e., when vocal explanations or directions are not necessary for the listeners to know what the signal means and what they should do).				
b. A specific point in time (that has no absolute value) is to be indicated (e.g., when the sound of a gong tells the listeners that something has happened or is about to happen, that they should be prepared for a message, etc.).				
c. A spoken message would compromise the security of a situation (i.e., when a coded tonal signal would be unrecognizable to persons not privy to the code).				
d. Noise conditions are unfavorable for receiving spoken messages.				
e. Speech channels are overloaded.				

	Yes	No	N/A	Comments
f. A spoken message could annoy listeners for whom it is not intended or when the spoken message could mask other messages.				
g. The intended listeners are familiar with the tonal signal implication or the tonal signal code.				
h. It is desired to use the simplest audio signal.				
2. A **spoken message** should be used when				
a. More message flexibility is needed than a tonal signal can convey.				
b. It is necessary to identify the source of the information.				
c. Listeners have not had training in a special tonal signal code.				
d. There is a need for rapid two-way exchanges of information.				
e. The intended information deals with a future time and when preparation is required (e.g., preparatory to initiating some operation during a countdown).				
f. Use of a tonal signal countdown could result in a miscount.				
g. Operational stress surrounding the intended listeners could cause them to forget the meaning of a tonal signal code.				

	Yes	No	N/A	Comments
h. The message is simple, short, and will not be referred to later.				
i. The message calls for immediate action, visual is already overburdened, or the job requires the user to move about continually.				
Criteria for Voice Signal Selection				
3. For auditory displays with voice output, different voices should be considered for use in distinguishing different categories of data.				
4. Verbal warning signals should consist of				
a. An initial alerting signal (non-speech) to attract attention and to designate the general problem.				
b. A brief, standardized speech signal (verbal message) which identifies the specific condition and suggests appropriate action.				
5. Verbal alarms for critical functions should be at least 20 dB above the speech interference level at the operating position of the intended receiver.				
6. The voice used in recording verbal warning signals should be distinctive and mature.				
7. Verbal signals should be presented in a formal, impersonal manner.				

	Yes	No	N/A	Comments
8. Verbal warning signals should be conditioned only when necessary to increase or preserve intelligibility.				
9. In selecting words to be used in audio warning signals, priority should be given to intelligibility, aptness, and conciseness, in that order.				
10. Computer speech outputs should be repeatable at user request.				
11. After each computer speech output, the computer should provide the user the choice of responding with wait, go ahead, or repeat.				
12. The user should be provided with a means of easily returning to the step in the program sequence immediately prior to the computer speech output.				
Audio–Visual				
<u>Supporting Information</u>				
1. Appropriate trend and status displays for minor upsets should be available to the user.				
2. Status displays for minor upsets should include a detailed alarm list that identifies alarms by name and number.				
<u>Alarm Handling Functions</u>				
3. Errors with a structured response pattern should be handled within the computer and should not trigger alarms.				

	Yes	No	N/A	Comments
4. Only those errors that require rapid operator decision-making and intervention should activate the redundant alarms.				
5. Early signs of a system going out of specification should be identified by low-level alarms since response time is not as critical.				
6. In some operations, it may be desirable to preprogram a hierarchy of alarms that could be altered if changes were made in manufacturing specifications.				
7. It should be possible to program the computer to display specific information each time one of the major system failures is detected and triggers an alarm.				
8. Information displayed to the operator should include the chronological order of failure.				
9. Information that will interfere with decision-making, such as alarms from systems that are secondarily affected by the initial problem, should not be presented to the operator unless requested.				
10. Numbers of alarms may be reduced by functional grouping.				
Feedback				
11. Once the operator has responded to an alarm with a controlling action, some				

	Yes	No	N/A	Comments
feedback should be indicated on the VDU to acknowledge that action has been taken.				
<u>Audio Signal Integration with Visual Displays</u>				
12. When used in conjunction with visual displays, audio warning devices should be supplementary or supportive. The audio signal should be used to alert and direct operator attention to the appropriate visual display.				
<u>Criteria for Audio/Visual Signal Selection</u>				
13. The frequency range of audio warning signals should be between 200 and 5000 Hz and, if possible, between 500 and 3000 Hz.				
14. If the user will ever be a considerable distance from the equipment in the performance of other tasks, the signal should be loud but of low frequency (less than 1000 Hz).				
15. If the user goes into another room or behind partitions, the signal should be of low frequency (below 500 Hz).				
16. If there is substantial background noise, the signal should be of a readily distinguishable frequency.				
17. If an auditory signal must attract attention and the above guidelines are inadequate, the signal should be modulated.				

	Yes	No	N/A	Comments
18. The frequency of a warning tone should be different from that of the electric power employed in the system, to preclude the possibility that a minor equipment failure may generate a spurious signal.				
19. The intensity, duration, and source location of audio alarms and signals should be compatible with the acoustical environment of the intended receiver as well as the requirements of other personnel in the signal areas.				
20. Audio warning signals should not be of such intensity as to cause discomfort or ringing in the ears as an aftereffect.				
Operator Control				
21. Controls for operator response to the annunciator system should include silence, acknowledge, reset, and test controls.				
22. The alarm should cease **only** after the user responds appropriately.				
23. It should be possible to silence an auditory alert signal from any set of annunciator response controls in the primary operating area.				
24. The acknowledgment control should terminate the flashing of a visual tile and have it continue at steady illumination until the alarm is cleared.				
25. Acknowledgment should be possible only at the work station where the alarm originated.				

	Yes	No	N/A	Comments
26. The reset control should silence any audible signal indicating clearance and should extinguish tile illumination.				
27. The reset control should be effective only at the workstation for the annunciator panel where the alarm initiated.				
28. The test control should actuate the audible signal and flashing illumination of all tiles in a panel.				
29. Periodic testing of annunciators should be required and controlled by administrative procedure.				
Workstation Layout				
30. To facilitate blind reaching, repetitive groups of annunciator controls should have the same arrangement and relative location at different workstations.				
Other Techniques				
1. Graphic coding methods should be used to present standardized qualitative information to the operator or to draw the operator's attention to a particular portion of the display.				
Screen Structures				
2. Extra spacing, horizontal and vertical lines of differing widths, and perhaps color should be used to set off and highlight data.				

	Yes	No	N/A	Comments
3. Special symbols (e.g., bullets or arrows) should be used to indicate position and to direct attention.				
4. Other methods of coding which should be considered for graphic displays and computer-generated drawings include motion, focus, distortion, and line orientation on the display surface.				
Borders				
5. A border should be used to improve the readability of a single block of numbers or letters.				
6. If several labels or messages are clustered in the same area, distinctive borders should be placed around the critical ones only.				
Spacing				
7. When a special symbol is used to mark a word, it should be separated from the beginning of the word by a space.				
Line				
8. Auxiliary methods of line coding should be considered for graphics applications, including variation in line type (solid, dashed, dotted) and width (boldness).				
9. When a line is added simply to mark or emphasize a displayed item, it should be placed **under** the designated item.				

	Yes	No	N/A	Comments
Special Applications				
10. Visual dimensions that should be considered for special display coding applications include variation in texture, focus, and motion.				
Use of Colors to Highlight Data				
11. Related data which are distributed about the screen and data to be up-dated, etc. should be highlighted in white.				
DYNAMIC DISPLAY				
Display Motion				
1. The speed of a graphic showing fluid flow in a pipe should be greater than 7.28 mm/s (0.29 in./s) but less than 295 mm/s (11.8 in./s).				
2. Changing values which the operator uses to identify rate of change or to read gross values should not be up-dated faster than 5 s nor slower than 2 s when the display is to be considered as real time.				
3. A display freeze mode should be provided to allow close scrutiny of any selected frame.				
4. Display formats should be designed to optimize information transfer to the operator by means of information coding, grouping, and appropriate information density.				

	Yes	No	N/A	Comments
5. Rate of motion should not exceed 60 deg/s of visual angle change with 20 deg/s preferred.				
Digital Counters				
1. Numerals should not follow each other faster than 2/s when the operator is expected to read the numerals consecutively.				
2. Changing digital values which the operator must reliably read should not be updated faster than 1/s, with a 2/s minimum time preferred.				
INFORMATION FORMATS				
Analog				
1. Analog displays should not be used when quick, accurate readings are a criterion.				
2. Numbers should increase clockwise, left to right, or bottom to top, depending on the display design and orientation.				
3. For one-revolution, circular scales, zero should be at 7 o'clock and the maximum value should be at 5 o'clock, with a 10-degree break in the arc.				
4. When check-reading positive and negative values, the zero or null position should be at 12 o'clock or 9 o'clock.				
5. All numbers should be oriented upright.				

	Yes	No	N/A	Comments
6. Numbers should be outside the graduation (tick) marks unless doing so would constrict the scale.				
7. The pointer on fixed scales should extend from the right of vertical scales and from the bottom of horizontal scales.				
8. The pointer on fixed scales should extend to but not obscure the shortest graduation marks.				
9. Graduation interval values of fixed scales should be 1, 2, 5, or decimal multiples thereof. Numbering by 1, 10, or 100 is recommended for progressions.				
10. Nine should be the maximum number of tick marks between numbers.				
11. Tick marks should be separated by at least 0.07 inches for a viewing distance of 28 inches (710 mm) under low illumination (0.03-1.0 fL).				
12. Dials should not be cluttered with more marks than necessary for precision.				
13. Zones should be colorcoded by edge lines or wedges. Red, yellow, and green should be used.				
14. Shape coding or striping should be used when red lighting or blackout station conditions prevail.				
15. Information should be in a directly usable form (for example, percent, RPM).				

	Yes	No	N/A	Comments
16. Fixed-scale, moving-pointer displays should be used rather than moving-scale, fixed-pointer displays.				
Digital				
1. Each digital display should have a label to identify its meaning.				
2. Digital displays should include the appropriate number of significant figures for the required level of accuracy.				
3. Digital displays should accommodate the full range of the variable (i.e., highest and lowest values).				
4. Digital displays should change slowly enough to be readable.				
5. Digital displays should be provided with arrows to indicate the direction of change (if that is likely to be needed).				
6. If more than four digits are required, they should be grouped and the groupings separated as appropriate by commas, a decimal point, or additional space.				
7. Multidigit counters should be oriented to read horizontally from left to right.				
8. Simple character fonts should be used.				
9. Horizontal spacing between numerals should be between one-quarter and one-half the numeral width.				

	Yes	No	N/A	Comments
Binary Indicator				
1. Binary indicators should be clearly labeled and understood.				
2. For quantitative measurements, binary indicators should be used only for check-reading purposes.				
3. Where meaning is not apparent, labeling should be provided close to the status indicator.				
4. When monochrome is not used, The color of the indicator should be clearly identifiable.				
5. Symbolic legends should be clear and unambiguous as to their meaning.				
6. Legend text should be short, concise, and unambiguous.				
7. Legend nomenclature and abbreviations should be standard and consistent with usage throughout the control room and in the procedures.				
8. Legends should be worded to tell the status indicated by the display.				
9. The legends of illuminated indicators should be readily distinguishable from legend push buttons by form, size, or other factors.				
Bar/Column Charts				
1. Each bar on the display should have a unique identification label.				

	Yes	No	N/A	Comments
2. Bar charts should contain reference(s) to the normal operating condition(s).				
3. Column charts should be used when the direction of change of the measurement is to be emphasized or when time is represented by one of the axes of the chart.				
4. Stroke type charts are alternatives to conventional full bars.				
Band Chart				
1. All items on a band chart should be related to the total.				
Linear Profile				
1. A horizontal line representing normal operating conditions should be superimposed on the display.				
2. The area below the profile line should be shaded to provide a more distinguishable profile.				
3. Labels should be provided along the bottom to identify each parameter.				
4. Linear profile charts should be used in applications where detection of abnormal events is important.				
Circular Profile				
1. The chart should be designed so that it forms recognizable geometric patterns for specific abnormal conditions.				

	Yes	No	N/A	Comments
2. Labels should be provided to identify each radial line.				
3. The area within the profile should be shaded to enhance the operator's perception of plant status.				
Single Value Line Chart				
1. The target area should be defined.				
2. Old data points should be removed after some fixed period of time.				
Trend Plot				
1. If time is plotted on the X axis, it should increase from left to right; if time is plotted on the Y axis, it should increase from top to bottom.				
2. When more than one parameter is presented in a plot, there should be means of identifying each individual parameter.				
3. When more than one parameter is displayed on a plot, the grouping of the parameters should enhance the operator's assessment of the safety status of the plant.				
4. CRT-displayed trend plot scales should be consistent with the intended functional use of the data.				
5. Graphic lines should contain a minimum of 50 resolution lines per inch.				

	Yes	No	N/A	Comments
6. Trend displays should be capable of showing data collected during time intervals of different lengths.				
7. If the general shape of the function is important in making decisions, a graph should be chosen rather than a table or scale; if interpolations are necessary, graphs and scales should be used in preference to tables.				
8. Graphs should be constructed so that numbered grids are bolder than unnumbered grids. If ten-grid intervals are used, the fifth intermediate grid should be less bold than the unnumbered grids.				
9. For tasks requiring both time to estimate trends and accuracy, the line graph should be used rather than horizontal bar or column charts.				
10. Time history displays of safety status parameters should present the 30-min interval immediately preceding current real time.				
Mimic Display				
1. A mimic should contain just the minimum amount of detail required to yield a meaningful pictorial representation.				
2. Abstract symbols should conform to common electrical and mechanical symbol conventions whenever possible.				

	Yes	No	N/A	Comments
3. Differential line widths should be used to code flow paths (e.g., significance, volume, level).				
4. Mimic lines should not overlap.				
5. Flow directions should be clearly indicated by distinctive arrowheads.				
6. All mimic origin points should be labeled or begin at labeled components.				
7. All mimic destination or terminal points should be labeled or end at labeled components.				
8. Component representations on mimic lines should be identified.				
9. Symbols should be used consistently.				
CONTROLS AND INPUT DEVICES				
Keyboard Layout				
Keystroke Feedback				
1. An indication of control activation should be provided (e.g., snap feel, audible click, or associated or integral light).				
Key Actuation Force				
1. The force required for key displacement should be 0.25 to 1.5 N.				
2. The force required for key displacement should be 0.3 to 0.75 N for repetitive keying tasks.				

ESSENTIAL USER-COMPUTER GUIDELINES

	Yes	No	N/A	Comments
Key Rollover				
1. N-key rollover capability should be implemented for the reduction of keying errors.				
Key Travel (Displacement)				
1. Key displacement should be 0.03 to 0.19 in. for numeric keys and 0.05 to 0.25 in. for alphanumeric keys.				
2. Displacement variability between keys should be minimized.				
Key Color/Labeling				
1. All controls should be appropriately and clearly labeled in the simplest and most direct manner possible.				
2. Functional highlighting of the various key groups should be accomplished through the use of color-coding techniques.				
3. Key symbols should be etched to resist wear and colored with high contrast lettering.				
4. Color of alphanumeric keys should be neutral (e.g., beige, grey) rather than black or white or one of the spectral colors (red, yellow, green, or blue.)				
5. Keys should be matt finished.				
6. Keys should be labeled with a nonstylized font.				

ESSENTIAL USER-COMPUTER GUIDELINES

	Yes	No	N/A	Comments
Key Dimension/Spacing				
1. The linear dimensions of the key tops should be from 0.385 to 0.75 in., with 0.5 in. preferred.				
2. Separation between adjacent key tops should be 0.25 inch.				
3. Push-button height for decimal entry keypads should be from 1/4 to 3/8 inch.				
4. Key height for alphanumeric keyboards should be from 3/8 to 1/2 inch.				
Keyboard Slope				
1. Keyboards should have a slope of 15 to 25 degrees from the horizontal, with 12 to 18 degrees preferred.				
2. The keyboard slope should be adjustable.				
Keyboard Thickness				
1. The thickness of the keyboard, i.e., base to the home row of keys, should be less than 50 mm (acceptable) with 30 mm or less preferred.				
Special Function Keys				
1. When dedicated controls are used to initiate/activate functions, the keys should be grouped together.				
2. Function controls should be easily distinguished from other types of keys on the computer console.				

	Yes	No	N/A	Comments
3. Each function control should be clearly labeled to indicate its function to the operator.				
4. When function keys are included with an alphanumeric keyboard, the function keys should be physically separate.				
5. Keys with major or fatal effects should be located so that inadvertent operation is unlikely.				
Soft Programmable Keys				
1. Commands should be consistent throughout PDP procedures.				
2. Use blink coding when there is an urgent need for the subject's attention.				
3. The system should allow users to step backward or forward through menus or procedures.				
4. PDPs should not be used in complex applications such as the sole display and control, e.g., use in conjunction with CRT.				
5. PDPs may be used as the sole device with simple applications such as camera control.				
6. PDPs should contain abbreviations which are easily recognized by the **user**. (In many cases there is a six letter limit on a button for labels.)				

	Yes	No	N/A	Comments
Numeric Keypad				
1. Terminals which are often used as calculators should be provided with an auxiliary numeric key set.				
2. The configuration of a keyboard used to enter solely numeric information should be a 3 x 3 x 1 matrix with the zero digit centered on the bottom row.				
3. The layout of keyboard numeric pads should be either telephone or calculator style.				
ALTERNATE INPUT DEVICES				
Light Pens				
1. Light pens should be used for cursor placement, text selection, and command construction.				
2. Tasks involving light pens should not require frequent, alternating use of the light pen and the keyboard.				
3. Tasks involving light pens should not require long, continuous intervals of light pen use.				
4. The light pen should be 12 to 18 cm (4.7 to 5.1 in.) long and 0.7 to 2 cm (0.3 to 0.8 in.) in diameter.				
5. Convenient clips should be provided at the lower right side of the CRT to hold the pen when it is not in use.				

	Yes	No	N/A	Comments
6. Movement of the pen in any direction on the screen should result in smooth movement of the follower in the same direction.				
7. Discrete placement of the stylus at any point on the screen should cause the follower to appear at that point and remain steady in position so long as the pen is not moved.				
8. Refresh rate for the follower should be sufficiently high to ensure the appearance of a continuous track whenever the pen is used for generation of free-drawn graphics.				
Joysticks				
General				
1. Joystick controls should be used for tasks that require precise or continuous control in two or more related dimensions.				
2. In rate-control applications which allow the follower to transit beyond the edge of the display, indicators should be provided to aid the operator in bringing the follower back onto the display.				
Isotonic Joysticks				
3. Isotonic joysticks which are used for rate control should be spring-loaded for return to center when the hand is removed.				

	Yes	No	N/A	Comments
4. Isotonic joysticks should not be used in connection with automatic sequencing of a CRT follower unless they are instrumented for null return or are zero-set to the instantaneous position of the stick at the time of sequencing. Upon termination of the automatic sequencing routine, joystick center should again be registered to scope center.				
5. Isotonic/displacement joysticks should be 1/4 to 5/8 inch in diameter and 3 to 6 inches long.				
6. Resistance force of the joystick should be 12 to 32 ounces.				
7. Full displacement of the joystick should not exceed 45 degrees.				
8. Isotonic/displacement joysticks should be provided with the following clearances: a. Display to stick—15-3/4 in. b. Around stick—4 in. c. Stick to shelf front—4-3/4 in. to 9-7/8 in.				
9. Movement should be smooth in all directions, and rapid positioning of the follower on the display should be attainable without noticeable backlash, cross-coupling, or need for multiple corrective movements.				
10. Control ratios, friction, and inertia should meet the dual requirements of rapid gross positioning and precise fine positioning.				

	Yes	No	N/A	Comments
11. Recessed mounting or pencil attachments may be utilized to provide greater precision of control.				
12. When used for generation of free-drawn graphics, the refresh rate for the follower on the CRT should be sufficiently high to ensure the appearance of a continuous track.				
13. Delay between control movement and the confirming display response should be minimized and should not exceed 0.1 s.				
14. When positioning accuracy is more critical than positioning speed, isotonic displacement joysticks should be selected over isometric joysticks.				
15. Isotonic displacement joysticks should be used for such functions as data pickoff and generation of free-drawn graphics.				
Isometric Joystick				
16. The isometric joystick should be used for such functions as data pickoff.				
17. Isometric joysticks should ordinarily not be used in any application where it would be necessary for the operator to maintain a constant force on the stick to generate a constant output over a sustained period of time.				
18. Finger-grasped isometric joysticks should comply with the same dimensional criteria as isotonic joysticks.				

	Yes	No	N/A	Comments
19. Hand-grasped isometric joysticks, when integral switching is required, should be between 4.3 to 7.1 inch long and have a maximum grip diameter of 2 inches.				
20. Hand-grasped isometric joysticks should have minimum clearances of 4 in. at the sides and 2 in. at the rear.				
21. Hand-grasped isometric joysticks should have a maximum resistance force of 26.7 lb for full output.				
22. The isometric stick should deflect minimally in response to applied force but may deflect perceptibly against a stop at full applied force.				
23. The X and Y output should be proportional to the magnitude of the applied force as perceived by the operator.				
Tracker Ball				
1. A ball control should be used for such tasks as data pickoff.				
2. In any application which tracker ball controls, do not allow the ball to drive the follower on the display off the edge of the display.				
3. When tracker ball controls are used to make precise or continuous adjustments, wrist support or arm support or both should be provided.				
4. Tracker ball controls should conform to the dimensions listed in Table 11.				

	Yes	No	N/A	Comments
5. The tracker ball control should be capable of rotation in any direction so as to generate any combination of X and Y output values.				
6. When moved in either the X or Y directions alone, there should be no apparent cross-coupling (follower movement in the orthogonal direction).				
7. While manipulating the control, neither backlash nor cross-coupling should be apparent to the operator.				
8. Control ratios and dynamic features should meet the dual requirement of rapid press positioning and smooth, precise fine positioning.				
9. Tracker balls should be used in graphic applications requiring position and selection.				
Grid and Stylus Devices				
1. Grid and stylus devices should be used for data pickoff, entry of points on a display, generation of free-drawn graphics, and similar control applications.				
2. Transparent grids which are used as display overlays should conform to the size of the display.				
3. Grids which are displaced from the display should approximate the display size and should be mounted below the display in an orientation to preserve directional relationships to the maxi-				

	Yes	No	N/A	Comments
mum extent (i.e., a vertical plane passing through the north/south axis on the grid shall pass through or be parallel to the north/south axis on the display).				
4. Movement of the stylus in any direction on the grid surface should result in smooth movement of the follower in the same direction.				
5. Discrete placement of the stylus at any point on the grid should cause the follower to appear at the corresponding coordinates and to remain steady in position so long as the stylus is not moved.				
6. Refresh rate for the follower should be sufficiently high to ensure the appearance of a continuous track whenever the stylus is used in generation of free-drawn graphics.				
X-Y Controller (Mouse)				
1. The mouse controller should be used for main item selection, scrolling, data retrieval, and data entry.				
2. The controller should have physical dimensions of 1.5 to 3 in. width, 3 to 5 in. length, and 1 to 2 in. thickness.				
3. The design of the controller and placement of the maneuvering surface should allow the operator to consistently orient the controller to within ±175 mrad (10°) of the correct orientation without visual reference to the controller.				

	Yes	No	N/A	Comments
4. The controller should be easily movable in any direction without a change of hand grasp and should result in smooth movement of the follower in the same direction $\pm$175 mrad (10°).				
5. The controller should be **cordless** and should be operable with either the left or right hand.				
6. A complete excursion of the controller from side to side of the maneuvering area should move the follower from side to side on the display regardless of scale setting or offset unless expanded movement is selected for an automatic sequencing mode of operation.				
7. Applications which allow the controller to drive the follower off the edge of the display should provide indicators to assist the operator in bringing the follower back onto the display.				
Automatic Speech Recognition				
1. Automatic speech recognition (voice input devices) should be limited to relatively simple input tasks.				
2. Keyboard is to be preferred for entry of numeric strings.				
3. Voice entry may have an acceptable error rate for entry of alphanumeric strings.				

ESSENTIAL USER-COMPUTER GUIDELINES

	Yes	No	N/A	Comments
Touch Screen Input				
1. Touch screens should be used for main item selection, scrolling, data retrieval, and data entry.				
2. The terminal should recognize a person's touch in approximately 100 ms.				
3. The system should accept only one command at a time, indicate that the command has been accepted, and respond in a time commensurate with the activity.				
4. The sensitive areas should be large enough to allow entry using fingers and allow for parallax due to CRT screen curvature.				
5. To avoid alteration of color codes, touch screens should be toned with a neutral tint.				
6. Touch screens are not recommended if task requires holding arm up to the screen for long periods of time.				
7. Use discriminable audible beeps (used to supply feedback) when more than one touch screen will be installed at more than one work station.				

ESSENTIAL USER-COMPUTER GUIDELINES

	Yes	No	N/A	Comments
CONTROL/DISPLAY INTEGRATION				
USER DIALOGUE				
Question and Answer				
1. Question-and-answer dialogue should be used primarily for routine data entry tasks where the user has little or no training.				
2. The data items should be known and their ordering constrained.				
3. The computer response should be moderately fast.				
Form Filling				
1. Form-filling dialogue should be used when some flexibility in data entry is needed, such as the inclusion of optional as well as required items, where users will have moderate training, and/or where computer response may be slow.				
Menus—Option Codes/Applications				
1. Selection should be accomplished by keyed entry of corresponding codes or by other means such as programmed multifunction keys labeled in the display margin.				
2. When menu selection is accomplished by code, that code should be keyed into a standard command entry area (window) in a fixed location on all displays.				

	Yes	No	N/A	Comments
3. When control entries will be selected from a discrete set of options, those options should be displayed at the time of selection.				
4. Displayed options should be worded in terms of recognized commands or command elements.				
5. If menu selections must be made by keyed codes, each code should be the initial letter (or letters) of the displayed option label rather than an arbitrary number.				
6. If letter codes are used, those codes should be used consistently in designating options at different steps in a transaction sequence.				
7. Menus should be used to minimize training needs.				
8. Menus should be used when users have little or no typing skills.				
9. Menus should be used when the system has a limited keyboard.				
Menus—Option Selection				
1. Computer response should be fast.				
2. Each menu display should require just one selection by the user.				
3. Displayed menu options should be listed in a logical order; if no logical structure is apparent, then options should be displayed in order of their				

	Yes	No	N/A	Comments
expected frequency of use, with the most frequent listed first.				
4. Displayed menu lists should be formatted to indicate the hierarchic structure of logically related groups of options, rather than as an undifferentiated string of alternatives.				
5. If menu options are grouped in logical sub-units, those groups should be displayed in order of their expected frequency of use.				
6. Use the same color for menus within the same group.				
Menu Navigation				
1. When hierarchic menus are used, the user should be given some displayed indication of current position in the menu structure.				
2. When hierarchic menus are used, a single key action should permit the user to return to the next higher level.				
3. Menus provided in different displays should be designed so that option lists are consistent in terminology and ordering.				
4. Experienced users should be provided means to bypass a series of menu selections and make an equivalent command entry directly.				
5. When a user can anticipate menu selections before they are presented,				

	Yes	No	N/A	Comments
means should be provided to enter several stacked selections at one time.				
6. Menu displays for a system still under development might indicate future options not yet implemented, but those options should be specially designated in some way.				
Command Language				
1. When command language is used for control entry, an appropriate entry area should be provided in a consistent location on every display, preferably at the bottom.				
2. The words chosen for a command language should reflect the user's point of view and not the programmer's.				
3. Abbreviation of entered commands (i.e., entry of the first 1 to 3 letters) should be permitted to facilitate entry by experienced users.				
4. Do **not** label two commands DISPLAY and VIEW when one permits editing displayed material and one does not.				
5. The user should be able to request display of a file by name alone without having to enter any further information such as file location in computer storage.				
6. The user should be able to request prompts as necessary to determine required parameters in a command entry or to determine available options for an appropriate next command entry.				

	Yes	No	N/A	Comments
7. The user should be able to enter commands without punctuation.				
8. Neither the user nor the computer program should have to distinguish between single and multiple blanks in a command entry.				
9. The computer should be programmed to recognize common misspellings of commands and to display inferred correct commands for user confirmation rather than requiring reentry.				
10. When a command entry is not recognized, the computer should initiate a clarification dialogue rather than rejecting the command outright.				
11. The system shall accept user input without discriminating between upper and lower case.				
12. Command language assumes highly experienced users.				
13. Command language assumes trained users.				
Query Language				
1. Query language dialogue should be used as a specialized subcategory of general command language for tasks emphasizing unpredictable information retrieval (as in many analysis and planning tasks).				
2. The organization of the query language should match the data structure perceived by users to be natural.				

	Yes	No	N/A	Comments
3. One single representation of the data organization should be established for use in query formulation, rather than multiple representations.				
4. The need for quantificational terms in query formulation should be minimized or eliminated.				
5. Use of operators subject to frequent semantic confusion, such as "or more" and "or less," should be minimized.				
Natural Language				
1. Consider using some constrained form of natural language in applications where task requirements are broad-ranging and poorly defined, where little user training can be provided, and where computer response will be fast.				
Expert Systems				
1. The system should make it easy and natural for a user to inquire about any details desired.				
Expert System Dialogue				
2. The system should support a flexible dialogue that permits either the user or the expert system to initiate an action or request for information without cancelling an ongoing transaction.				
3. The user-expert system dialogue should be flexible in terms of the type and sequencing of user input it will accept.				

	Yes	No	N/A	Comments
User-Assistance				
4. The system should be capable of supporting speculative analysis (e.g., what if scenarios) by providing a "reconnoiter mode" that allows the user to investigate the effects of an action without actually implementing the action.				
Problem Definition				
5. The knowledge required to perform all functions allocated to the expert system should be directly accessible by the expert system. Requirements for the expert system to query the user to obtain information for routine functions should be minimized.				
6. The capability for the user to supercede the current request for information from the expert system in order to input information related to a different transaction should be provided.				
Information Display				
7. The expert system should have the capability to graphically represent its rules network. This capability should be available to the user as an adjunct to the explanation subsystem.				
Consultation				
8. The expert system should automatically record all rules invoked during a consultation. Following a consultation, the explanation facility should be capable of recalling each invoked rule				

	Yes	No	N/A	Comments
and associating it with a specific event (i.e., question or conclusion) to explain the rationale for the event.				
9. The expert system should be able to respond to user requests to clarify or restate questions and assertions.				
10. At any point during a transaction, the expert system should be able to explain which problem-solving strategy is being employed, why a particular strategy was selected, and the current status of the application.				
<u>Level of Explanation</u>				
11. The level of detail of information presented as part of an explanation or justification should be under the control of the user. As a minimum, the user should be able to specify three levels of detail: rules only, brief explanations, and detailed explanations.				
System Feedback				
Display Update Rate				
1. Update rates for continuous, real-time tracking tasks should not exceed 0.5 s.				
2. In general, update rates should not exceed 3 s.				
Response Time				
1. Response times should be within the maximums shown in Table 13.				

	Yes	No	N/A	Comments
2. Response time deviations should be less than one-half the mean response time.				
System Status Indication				
1. An indication that the computer or control panel is functioning normally should be provided on the CRT display.				
Routine Status Information				
1. When system functioning requires the operator to stand by, periodic feedback should be provided to the operator to indicate normal system operation and the reason for the delay.				
2. When a process or sequence is completed by the system, positive indication should be presented to the operator concerning the outcome of the process and requirements for subsequent operator actions.				
3. If at any time the keyboard is locked or the terminal is otherwise disabled, that condition should be signaled by disappearance of the cursor from the display and (especially if infrequent) by some more specific indicator such as an auditory signal.				
4. Status information should be available indicating current load (multiple users assumed) and/or current system performance.				

	Yes	No	N/A	Comments
5. Relevant status information for external systems should be available to the user.				
6. When time tagging information is important, date-time signals should be available to users as an annotation on displays.				
7. Status information should be available concerning the current status of alarm settings, in terms of dimensions/ variables covered and values/catego- ries established as critical.				
8. Every user input should consistently produce some perceptible response output from the computer.				
9. Computer response to user entries should be rapid, with consistent timing as appropriate to different types of transactions.				
10. Following user interrupt of data proc- essing, an advisory message should be displayed assuring the user that the system has returned to its previous status.				
Performance/Job Aids				
1. Specific user guidance information should be available for display at any point in a transaction sequence.				
2. To serve as a home base or consistent starting point at the beginning of a transaction sequence, a general menu of control options should always be available for user selection.				

ESSENTIAL USER-COMPUTER GUIDELINES

	Yes	No	N/A	Comments
3. Hierarchic menus should be organized and labeled to guide the user within the hierarchic structure.				
4. Control options that are generally available at any step in a transaction sequence should be treated as **implicit** options, i.e., need not be included in a display of step-specific options.				
5. The computer should be programmed to provide prompting, i.e., to display advisory messages to guide users in entering required data and/or command parameters.				
6. When users vary in experience (which is often the case), prompting should be an optional guidance feature that can be selected by novice users but can be omitted by experienced users.				
7. When the results of a user entry are contingent upon context established by previous entries, some indication of that context should be displayed to the user.				
8. Implicit cues for data entry should be provided by consistent and distinctive formatting of data fields.				
9. Following computer generation of display output, the cursor should automatically be positioned on the display in a location consistent with the type of transaction.				

ESSENTIAL USER-COMPUTER GUIDELINES

	Yes	No	N/A	Comments
10. Reference material should be available for on-line display to the user describing system capabilities and procedures.				
11. In applications where a user may employ command entry, the computer should provide an on-line command index to help guide user selection and composition of commands.				
12. A complete dictionary of abbreviations used for data entry, data display, and command entry should be available for on-line user reference and in system documentation.				
13. When codes are assigned special meaning in a display, a definition should be provided at the bottom of the display.				
14. In system applications where it is warranted, the user should be able to request a displayed record of past transactions in order to review prior actions.				
15. In addition to explicit aids (labels, advisory messages) and implicit aids (cueing) provided in user interface design, there should also be a capability for a user to request further on-line guidance by a request for HELP.				
16. When an initial HELP display provides only summary information, more detailed explanations should be available in response to repeated user requests for HELP.				

	Yes	No	N/A	Comments
17. Novice users should be able to browse on-line HELP displays, just like a printed manual, to gain familiarity with system functions and operating procedures.				
18. For many system applications, an on-line training capability should be provided to introduce new users to system capabilities and to permit simulated hands-on experience in data handling tasks.				
SOFTWARE SECURITY				
Data Security				
1. Data security should be protected by automatic measures whenever possible, rather than by administrative procedures.				
2. User interface design should provide consistent procedures for data transactions, including data entry and error correction, data change, and deletion.				
3. Inputs to the computer, including data entries and control entries, should require explicit user actions.				
4. When the result of user action is contingent upon prior selection among differently defined operational modes, mode selection should be continuously indicated to the user, particularly when user inputs in that mode might result in unintended data loss.				

ESSENTIAL USER-COMPUTER GUIDELINES

	Yes	No	N/A	Comments
5. User interface design should deal appropriately with all possible control entries, correct and incorrect, without introducing unwanted data change.				
6. For both data entry and control entry, the user should be able to edit composed material before initial entry and also before any required reentry.				
7. For both data entry and control entry, the user should be required to resolve any detected ambiguity requiring computer interpretation.				
8. The user should be warned of potential threats to data security by appropriate messages and/or alarm signals.				
Policy-Related Issues				
9. Computer security procedures should be understood by all staff.				
10. Computer security policies should be strongly supported by management.				
11. Design documentation explicitly delegates controls to be used.				
12. Policy is established whereby employees do **not discuss** security procedures outside of the job environment.				
13. Unbeknownst to the user, the computer automatically logs user ID and keeps record of file access and work performed.				

ESSENTIAL USER-COMPUTER GUIDELINES

	Yes	No	N/A	Comments
14. The system is kept free of "Shareware" and other programs which may contain viruses.				
<u>Physical Access</u>				
15. Personnel are conspicuous by virtue of the fact that they are required to wear a badge.				
16. Visitors are required to wear identification.				
17. Passwords are employed by all users.				
18. Passwords are changed every 90 days.				
19. Passwords are changed every two weeks (high security access).				
20. Physical key locks are provided.				
<u>Control Implementation</u>				
21. Internal and external security audits are conducted on a regular basis.				
22. Commensurate with review of security, reliability of the system as a whole should be calculated.				
23. Data conversion procedures are subject to scrutiny.				
24. There are vertical controls, e.g., those between levels of the organization.				
25. There are horizontal controls, those between departments or agencies.				

ESSENTIAL USER-COMPUTER GUIDELINES

	Yes	No	N/A	Comments
26. Standards are in place which call for the use of controls.				
27. Construction practices should promote a fireproof and waterproof environment.				
WORKPLACE LAYOUT				
ANTHROPEMETRICS				
Keyboard Base				
1. The keyboard base height for a seated workplace should be from 56 to 77 cm (22 to 30 in.).				
2. The keyboard base height for a standing or a sitting/standing workplace should be from 90 to 93 cm (35.5 to 36.5 in.).				
Working Level				
1. Working level height for a sitting workplace should be from 66 to 81 cm (26 to 32 in.).				
2. Working level height for a standing workplace should be from 90 to 107 cm (35.5 to 42 in.).				
3. Working level height for a sitting/standing workplace should be from 90 to 102 cm (35.5 to 40 in.).				
4. Working level width should be from 61 to 76.5 cm (24.4 in to 30.6 in), 76.5 cm preferred.				

	Yes	No	N/A	Comments
5. Working level depth should be from 41 to 64 cm (16.4 in. to 25.6 in.), 64 cm preferred.				
Keyboard Home Row				
1. Keyboard home row height should be 66 to 78 cm (26 to 30.5 in.).				
Screen				
1. The screen height for a seated workplace should be from 15 to 117 cm (6 to 46 in.), with 99 cm (39 in.) preferred.				
2. The screen height for a standing workplace should be from 104 to 178 cm (41 to 70 in.).				
3. The screen viewing angle should be within 35 degrees of the horizontal line of sight, with about 15 degrees below the horizontal line of sight preferred.				
Viewing Distance				
1. The viewing distance should be 33 to 80 cm (13 to 30 in.), with 46 to 61 cm (18 to 24 in.) preferred.				
Footrest				
1. The footrest should be 18 in. below the level of the seat and should be adjustable in 2-in. increments of height.				
2. Rectangular footrests should be 30 cm (12 in.) deep by 41 cm (16 in.) wide.				

	Yes	No	N/A	Comments
3. Circular footrests should have a diameter of 18 in.				
4. The footrest should be circular if it is part of the chair.				
Reach Envelope				
1. The functional reach envelope should be from 64 to 88 cm (25.2 to 34.6 in.).				
Position and Movement of the Head				
1. The normal inclination angle of the head should be from 16 to 22 degrees.				
2. A document holder should be provided to reduce head movement while keying data from a document.				
Leg, Knee, and Foot Room				
1. For a sitting workplace, the following clearances should be provided:				
a. knee clearance (depth) - 46 to 51 cm (18 to 20 in.)				
b. leg clearance (depth) - 100 cm (39 in.)				
c. leg clearance (width) - 51 cm (20 in.)				
2. For a standing workplace, the following clearances should be provided:				
a. knee clearance (depth) - 10 to 50 cm (4 to 20 in.)				
b. foot clearance (depth) - 10 to 13 cm (4 to 5 in.)				

	Yes	No	N/A	Comments
c. foot clearance (height) - 10 to 12 cm (4 to 5 in.)				
d. foot clearance (width) - 51 to 70 cm (20 to 28 in.)				
Screen Orientation				
1. Screen orientation should be no greater than 45 degrees away from or toward the operator, with 15 degrees away from the operator preferred.				
2. Screen orientation should be adjustable.				
Chair				
1. The chair design should allow the user to maintain the following posture: knees flexed at an angle $\geq 90°$, elbows flexed at an angle $\geq 90°$, and torso at an angle slightly greater than 90° (100° to 155°).				
2. Seat height should be adjustable from 35 to 55 cm (14 to 22 in.).				
3. When the chair is provided with a footrest, it should be adjustable from 51 to 76 cm (20 to 30 in.), with the footrest a constant 46 cm (18 in.) below the seat.				
4. Seat width should be 43 to 51 cm (17 to 20 in.).				
5. Seat depth should be 38 to 46 cm (15 to 18 in.).				

	Yes	No	N/A	Comments
6. Backrest height should be 15 to 23 cm (6 to 9 in.).				
7. Backrest width should be 30 to 36 cm (12 to 14 in.).				
8. The seat cushion should be at least 1-in. thick.				
9. The armrests should be 5 cm (2 in.) wide, 20 cm (8 in.) long, and 19 to 28 cm (7.5 to 11 in.) above the compressed sitting surface. (Swing-away if appropriate.)				
Hardcopy Printer				
Hardware and Physical Characteristics				
1. Hard-finish, matt paper should be used to avoid smudged copy and glare.				
2. There should be a positive indication of the remaining supply of recording materials.				
3. Instructions for reloading paper, ribbon, ink, etc. should appear on an instruction plate attached to the printer.				
4. Printers should be part of the process computer system and be located in the primary operating area.				
5. Control room printers should provide the capability to record alarm data, trend data, and plant status data.				

	Yes	No	N/A	Comments
6. The system should, if possible, be designed to provide hard copy of any page appearing on the CRT at the request of the operator.				
7. Printer operation should not alter screen content.				
<u>Operator Information</u>				
8. A takeup device for printed materials should be provided which requires little or no operator attention and which has a capacity at least equal to the feed supply.				
9. It should be possible to annotate the print copy while it is still in the machine.				
10. The operator should always be able to read the most recently printed line.				
11. Printed material should have an adequate contrast ratio to ensure easy operator reading.				
12. When the printer is down during reloading, data and information which would normally be printed must not be lost.				
13. The recorded matter should not be obscured, masked, or otherwise hidden in a manner which prevents direct reading of the material.				
14. If the copy will be printed remote to the operator, a print confirmation or denial message should be displayed.				

ESSENTIAL USER-COMPUTER GUIDELINES

	Yes	No	N/A	Comments
15. Printed information should be presented in a directly usable form with minimal requirements for decoding, transposing, and interpolating.				
16. Printer used for recording trend data, computer alarms, and critical status information should have a high-speed printing capability of at least 300 lines a minute to permit printer output to keep up with computer output.				
Health and Safety				
1. The VDT should be provided with implosion safeguards.				
2. No fans, voltage, gears, or belts should be accessible to the user's fingers or body.				
3. No parts hotter than 140°F should be accessible to the user.				
4. Physical barriers and warnings should be provided for all live parts.				
5. Barriers should prevent hardware and small items from falling into areas of high voltage.				
6. All exposed corners and sharp edges should be smooth and rounded.				
7. Reviews of the literature (on radiation safety) and new surveys to attempt to measure radiation led to the conclusion that there is no radiation hazard for VDT operators and further routine surveys are not necessary.				

ESSENTIAL USER-COMPUTER GUIDELINES

	Yes	No	N/A	Comments
ENVIRONMENTAL FACTORS				
Background Noise				
1. Ambient noise level should be less than 70 dB(A) [less than 65 dB(A) preferred] in routine task areas and less than 55 dB(A) in task areas requiring a high level of concentration.				
2. Ambient noise should be free from high frequency tones (8000 Hz) and external (or extraneous), high-noise-level equipment.				
Temperature and Humidity				
1. Ambient temperature should be maintained from 18° to 29.5° K (65° to 85° F), with 21° to 27° K (70° to 80° F) preferred.				
2. Relative humidity should be from 20 to 60%.				
3. There should be no more than a 10° F difference between head and floor level.				
Lighting				
1. Workplace illuminance should be from 92 to 927 lx (adjustable), with a mean of 240 lx during the day shift and a mean of 184 lx during night shift.				
2. Emergency lighting level should be from 10 to 50 lx.				

ESSENTIAL USER-COMPUTER GUIDELINES

	Yes	No	N/A	Comments
Ventilation				
1. Air should be introduced at a minimum rate of 15 cfm per occupant, approximately 2/3 of which should be outside air that is filtered to remove hazardous or irritating particles.				
2. Air velocity should not exceed 45 fpm measured at operator head level and should not produce a noticeable draft.				
Static Electricity				
1. Relative humidity should be maintained at 40 ± 10% and should not be allowed to fall below 20%.				
2. An earth line should be provided between each VDT and the main system earth connection and carpeting material with a copper wire interweave.				

Index

W